KT-498-311

Introduction

This student book covers the foundation tier of the new OCR Gateway Science specification. The first examinations are in January 2007. It has been written to support you as you study for the OCR Gateway Science GCSE.

This book has been written by examiners who are also teachers and who have been involved in the development of the new specification. It is supported by other material produced by Heinemann, including online teacher resource sheets and interactive learning software with exciting video clips, games and activities.

Part of this new GCSE is assessments of your work that will be carried out by your school. These are called 'Science in the News' and 'Can Do Tasks', and are explained fully on pages 218–221.

We hope this book will help you achieve the best you can in your GCSE core Science Award and help you understand how much science affects our everyday lives. As citizens of the 21st century you need to be informed about science issues such as nuclear power, genetic engineering and pollution. Then you can read newspapers or watch television programmes and really have views about things that affect you and your family.

The following two pages explain the special features we have included in this book to help you to learn and understand the science, and to be able to use it in context. At the back of the book you will also find some useful tables, as well as a glossary and index.

About this book

This student book has been designed to make learning science fun. The book follows the layout of the OCR Gateway specification. It is divided into six sections that match the six modules in the specification with two for Biology, two for Chemistry and two for Physics: B1, B2, C1, C2, P1, P2.

The module introduction page at the start of a module (eg.below right) introduces what you are going to learn. It has some short introductory paragraphs, plus 'talking heads' with speech bubbles that raise questions about what is going to be covered.

Each module is then broken down into eight separate items (a–h), for example, B1a, B1b, B1c, B1d, B1e, B1f, B1g, B1h.

Each 'item' is covered in four book pages. These four pages are split into three pages covering the science content relevant to the item plus a 'Context' page which places the science content just covered into context, either by news-related articles or data tasks, or by examples of scientists at work, science in everyday life or science in the news.

Throughout these four pages there are clear explanations with diagrams and photos to illustrate the science being discussed. At the end of each module there are three pages of questions to test your knowledge and understanding of the module.

There are three pages of exam-style end of module questions for each module.

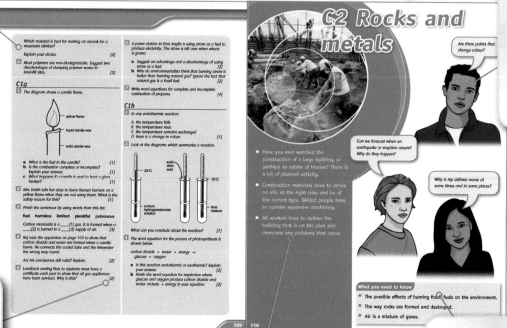

The talking heads on the module introductory page raise questions about what you are going to learn.

The numbers in square brackets give the marks for the question or part of the question.

The bulleted text introduces the module.

This box highlights what you need to know before you start the module.

Context pages link the science learned in the item with real life.

This box highlights what you will be learning about in this item.

General approach to the topic

Question box at the end.

Questions in the text make sure you have understood what you have just read.

Clear diagrams explain the science.

Some amazing facts have been included – science isn't just boring facts!

When a new word appears for the first time in the text, it will appear in bold type. All words in bold are listed with their meanings in the glossary at the back of

The keywords box lists all keywords in the item.

Drug safety

Stephanie is a scientist. She works for a big drug company. Her job is to discover new drugs and treatments to protect us from disease. Stephanie makes sure that all new medical treatments and new drugs are tested before they are used on human patients. This is to make sure that they are safe to use.

Some drugs are first tested on animals. They are then tested on human tissue in the laboratory. Some people object to testing drugs on animals and want to use other methods of testing such as computer models.

One computer model that Stephanie uses is the APACHE simulator. It is a complex computer database that she can use to predict how a drug will work before using it on real people. Some other scientists say that all drugs have side effects and it is much safer to test them on animals first.

This difference of opinion has caused a lot of conflict. Some animals' rights groups believe that it is all right to use violence to protect the laboratory animals from experimentation.

Even when drugs have been passed to use on patients, doctors have a form that they must fill in if a patient reports any adverse reactions to the drug. If there are enough adverse reactions the drug is withdrawn from use.

Questions

1. Suggest why it is important for Stephanie to discover new drugs.
2. Suggest why she must test the drugs before they are given to patients.
3. Explain how drugs can be tested using a computer model.
4. Explain what happens if a patient reports an adverse side effect from a new drug to their doctor.

14

Messages to the brain

In this item you will find out

- how the brain keeps in touch with all parts of the body
- about how the eye works
- how different parts of the body communicate with each other using neurones

Just imagine what it would be like if you lost your senses one by one – to become blind first, then deaf, then to have no sense of feeling and finally no sense of smell or taste. It's hard to imagine what it would be like. In experiments where volunteers are placed in an environment of sensory deprivation, they can usually only last for a few minutes before they press the 'panic button'.

Your body gathers information from the outside world in different ways. You get information from chemicals in food and air, from seeing and touching objects and from hearing sounds.

Your brain receives information about the outside world from your:

- eyes – vision
- skin – touch, pressure, heat and pain
- nose – smell
- tongue – taste
- ears – hearing and balance.

This vast amount of information is then sent along neurones to the brain.

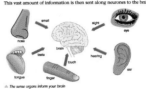

△ The sense organs inform your brain

Neurones are like telephone cables that send messages and instructions all round our bodies

Amazing fact

A human brain produces more electrical impulses in a single day than all of the world's telephone networks put together.

15

Nuclear radiation

Nuclear radiation is able to knock electrons out of atoms that it passes through. This **ionisation** of atoms produces charged particles. Nuclear radiation can be useful or it can be harmful. Some types of nuclear radiation can damage living cells and can cause cancer, but other forms of nuclear radiation can be used to treat cancer.

❓ What does the term ionisation mean?

Types of nuclear radiation

There are three types of nuclear radiation:

- **alpha** particles which can be stopped by a sheet of paper
- **beta** particles which need a few millimetres of metal to be stopped
- **gamma** rays which are partially absorbed by a few centimetres of metal.

alpha particle source — alpha particles are stopped by a sheet of paper

beta particle source — beta particles are stopped by a few millimetres of metal

gamma ray source — gamma rays need a few centimetres of metal to be absorbed

△ The penetrating powers of alpha, beta and gamma radiation

△ Thickness measurement with beta particles

❓ Name the three different types of nuclear radiation? Which is likely to be the most dangerous outside the body?

All three types of radiation are useful. Alpha particles are used in smoke detectors. The smoke alters the ionisation of the air caused by the alpha particles.

Beta particles are used in paper thickness gauges. A source shoots beta particles from strontium-90 at a continuous sheet of paper. If the thickness of the paper increases, then the number of beta particles reaching the detector decreases.

Doctors use beams of gamma rays from cobalt-60 to kill tumours inside the bodies of cancer sufferers, avoiding the need for surgery.

Medical equipment, such as scalpels and bandages, are sterilised by gamma rays. Items are wrapped in plastic and left close to some cobalt-60 for several hours. The gamma rays penetrate deep inside the items, killing any living organism which might cause infection.

Radiation all around

You are surrounded by radiation all the time. Sources of this background radiation include:

- cosmic rays, fast moving particles from outer space
- radon, a radioactive gas given off by soil and rocks in the ground
- your food, drink and other living things.

There is no way of escaping background radiation, but the risk it poses to your health is quite small.

△ Radiation can help to fight cancer

△ Tests show that sheep from some parts of the UK are still contaminated with radioactive fallout from the explosion of the Chernobyl reactor (in Russia) in 1986

❓ Name three different sources of background radiation.

Handling radioactive materials

Nuclear radiation is given out by radioactive materials. Radioactive materials can cause cancer, so they need to be handled carefully:

- keep them secure in shielded and clearly labelled storage
- handle them with tongs to keep them away from you
- wear gloves and protective clothing to stop any material getting on your skin
- keep exposure times short to minimise the risk.

❓ Explain why radioactive materials should be locked in labelled cupboards.

Amazing fact

The radioactive potassium in your bones emits about 10 000 beta particles every second.

△ Radioactive material must carry a hazard symbol

keywords

alpha • background radiation • beta • cosmic ray • gamma • ionisation • nuclear radiation • radioactive materials • radioactive waste

197

v

Contents

B1 Understanding ourselves

Someone told me the other day that I am what I eat.

What's that mean then?

I don't know.

It means if you put rubbish in your body, your body will be made out of rubbish.

Good job I don't eat out of dustbins then!

- This module is about understanding ourselves. It is only by understanding who and what we are that we can learn to make the right decisions to keep us healthy and, hopefully, have a long and happy life. In this module you will learn how we use food to obtain energy and what happens in parts of the world where food is scarce and people are starving. You will also learn how we are affected by disease and what we can do to keep healthy.

- One of the reasons why we have been so successful and are found living all over the world is that the human body can regulate and keep constant many of its internal processes, such as body temperature. But these internal mechanisms can be altered by drugs. Some people's lives have been ruined by drugs. You will learn about different types of drugs and the effect they can have on the body.

- Finally you will learn about DNA and how it makes us who and what we are. The study of DNA is an exciting new branch of science and it promises to bring about many new and wonderful changes in your lifetime.

What you need to know

- The types of food and what food contains, and that it provides us with energy.

- How we can stay fit and healthy.

- Microbes can cause disease.

- Variation exists in animals and plants and we are all different.

Fighting fit

In this item you will find out

- what causes blood pressure
- how the body gets its energy
- the difference between fitness and health

Have you ever wondered:

- Why we are still out of breath when we have stopped running?
- Why our muscles ache even after we have stopped exercising?
- Why, when your blood pressure is taken, the doctor gives you two readings instead of just one?

During exercise our heart rate increases. We also breathe more quickly and take in more air with each breath. This helps us to take in more oxygen through our lungs and pump that oxygen and **glucose** to the muscles where they are needed. We get the glucose from food in our gut. It also helps us to get waste carbon dioxide from the muscles to the lungs, where it can be breathed out.

It is important to have the right blood pressure. Too little and we feel faint and weak. Too high and it can cause long term damage such as a heart attack or a stroke. You will learn how life style choices affect our blood pressure. Most people with high blood pressure feel fine. It is often too late by the time they find out. This is why it is important, when you get older, that you get your blood pressure checked by your doctor on a regular basis.

▲ *'Your blood pressure is fine, 130 over 75'*

Amazing fact

The lungs of athletes can be up to 2dm³ larger and they take fewer breaths each minute than non-athletes.

a Explain why heart rate increases during exercise.

b Explain why breathing rate increases during exercise.

Aerobic respiration

In our body cells, glucose from food and oxygen from the air that we breathe in, react together to form carbon dioxide and water. This process is called **aerobic respiration** and it releases the **energy** that we need to live and exercise.

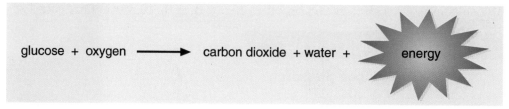

glucose + oxygen ⟶ carbon dioxide + water + energy

▲ Aerobic respiration

Anaerobic respiration

When we do vigorous exercise we need lots of energy. But sometimes our heart and lungs cannot provide our muscles with enough oxygen. When this happens, our cells carry out **anaerobic respiration** as well as aerobic respiration.

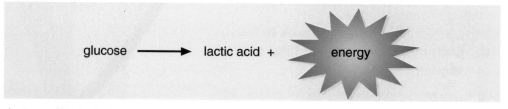

glucose ⟶ lactic acid + energy

▲ Anaerobic respiration

▲ Breathing and heart beat rate increase to deliver glucose and oxygen to the muscles

▲ Lactic acid builds up causing fatigue

The **lactic acid** builds up in our muscles. It causes muscle fatigue and makes them painful. Much less energy is produced during anaerobic respiration than during aerobic respiration.

c List one difference between aerobic and anaerobic respiration.

Under pressure

The heart is a muscular organ that pumps blood around the body. It pumps, or beats, about 80 times every minute. The blood in your arteries is under pressure all the time so it can reach all parts of your body. It is under pressure because of the contraction of your heart muscles.

When the heart contracts the blood pressure is at its highest. This is called the **systolic** pressure.

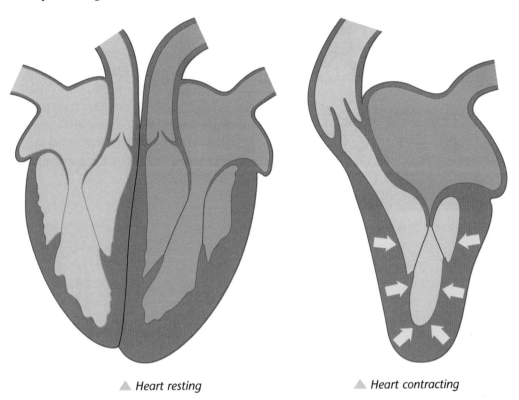

▲ Heart resting   ▲ Heart contracting

When the heart is resting the narrow blood vessels slow the blood flowing through them. This is when the blood is at its lowest pressure and is called the **diastolic** pressure.

Blood pressure is measured in units called mm Hg. A young fit person may have a blood pressure of about 120 mm Hg (systolic) over 70 mm Hg (diastolic).

Our blood pressure changes all of the time. It depends upon how active we are and our age, weight and lifestyle. It also depends on how much alcohol we drink and whether we are angry or calm.

As we get older our arteries get less elastic and clogged with fatty deposits.

Because the arteries are less elastic, they do not expand as much when the heart contracts. This causes the systolic pressure to be higher.

Because the arteries are clogged up, the diastolic pressure also rises when the heart is relaxing. This is what causes high blood pressure as we get older.

keywords
aerobic repiration • anaerobic respiration • diastolic • energy • glucose • lactic acid • systolic

Blood pressure	Systolic (mm Hg)	Diastolic (mm Hg)
normal	130	75
high	160	105
low	90	40

Measuring fitness

Deena is an athlete training for the next Olympics. She knows that fitness is not the same as being healthy and free from disease. Fitness is how efficiently her body can perform some of its functions. She is fit because she exercises.

▲ *This person is well but not fit*

▲ *This person is fit but not well*

When Deena is training, it is important for her trainer to know just how fit she is.

One way that her trainer can measure her fitness is to measure her breathing rate during exercise. As Deena exercises, her breathing rate speeds up. She also takes deeper breaths. Deena is connected to a machine that measures her breathing rate as she cycles. The graph shows what is happening to her breathing rate.

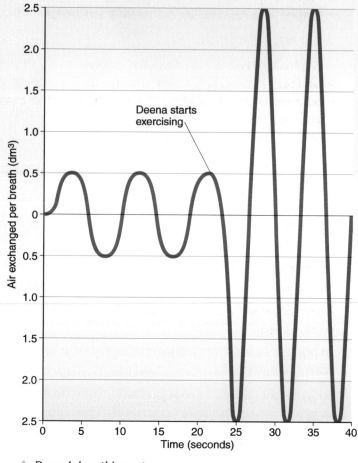

▲ *Deena's breathing rate*

Questions

1 State what happens to the number of breaths per minute when Deena exercises.

2 State what happens to the amount of air she takes in with each breath as she exercises.

3 Look at the graph. How much air is taken in with each breath during exercise?

4 Calculate how many breaths per minute Deena took before and after starting to exercise.

Diet dilemmas

In this item you will find out

- what is meant by a balanced diet

- how to calculate whether you are over or under weight

- how the digestive system works

All living organisms need food as a source of energy.

For humans, eating a balanced diet is very complicated because it is different for different people.

- Young people who are growing need to eat more than older people who have stopped growing.
- Active people need to eat more than people who are not active.

A balanced diet includes:

- **carbohydrates** and **fats** for energy
- **proteins** for growth and repair, and for energy when starving
- **minerals**, such as iron, to make haemoglobin and **vitamins**, such as vitamin C, to prevent scurvy
- fibre to prevent constipation
- water.

Carbohydrate includes all the starchy and sugary foods, such as bread, potatoes and rice, and sucrose, the sugar that you buy from the supermarket.

Fats include butter, margarine, and cooking oils such as olive oil and sunflower oil.

Proteins are found in foods such as meat, fish, milk and eggs, and in smaller amounts in any food that was once growing, e.g. plants.

Minerals, such as iron, are found in meat and also in some green plants such as broccoli.

Vitamins are found in fruit – especially vitamin C. However, they can also be found in the most unlikely places, such as potatoes.

Fibre is sometimes called roughage.

a Suggest what types of food a balanced diet should contain, and why it should contain roughage.

Minerals, vitamins, fibre and water don't provide us with any energy.

▲ *This is the amount of food eaten by one person in a day*

Amazing fact

We produce about 1.7 litres of saliva each day. That's about two lemonade bottles full.

What we get from food

Example of type of food	Food group	What we digest it into
Eggs	protein	amino acids
Butter	fat	fatty acids and glycerol
Potato	carbohydrate	simple sugars such as glucose

Animal proteins are called first class proteins because they contain all the amino acids we need. Most plant proteins do not contain all of the amino acids that we need, and are sometimes called second class proteins.

Problems with diets

Things can go wrong when we do not eat a balanced diet.

Some people eat too little because they are concerned about how they look and suffer from a condition called anorexia. Amino acids and simple sugars are absorbed into the blood stream. Fatty acids and glycerol are absorbed into the lymphatic system. Lymph is very similar to the watery liquid that oozes from grazed skin.

b Suggest how a vegetarian could get all of the amino acids that they need.

c Why is it more difficult for a vegetarian to have a balanced diet?

Kwashiorkor is a disease caused by not eating enough protein. In many parts of the world, usually in developing countries, people don't get enough protein in their diets.

Some people eat too much! When we get too fat we are called obese. Being obese increases our chances of getting arthritis, breast cancer, heart disease and diabetes.

Digestion

Food is digested physically as well as chemically.

In physical digestion the food is broken down into smaller chunks so that it can be swallowed and passed to the stomach. Chewing food in your mouth and the squeezing of food in your stomach are both examples of physical digestion. The food then passes into the small intestine where muscles contract and push the food along. They will even push the food along if you are standing on your head.

Chemical digestion in the stomach and small intestine involves large food molecules being broken down into smaller molecules by digestive **enzymes**. This is so the digested food molecules can be absorbed into your blood or your lymph.

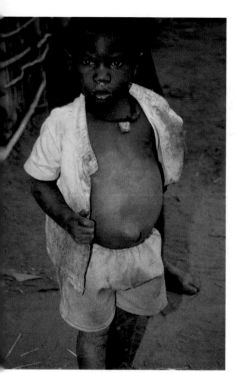
▲ A child with protein deficiency

d Explain the difference between physical and chemical digestion.

In the mouth

The teeth grind the food into much smaller pieces and the food is mixed with saliva. Saliva contains amylase, a **carbohydrase** enzyme, which breaks starch down into a simple sugar. This is why bread tastes sweet if you keep it in your mouth for about ten minutes.

In the stomach

Food is kept in the stomach for several hours. During this time, hydrochloric acid is added to the food. This kills most of the bacteria on the food and helps the enzymes to break it up into smaller molecules.

e Why is it important to destroy most of the bacteria found on our food?

Protease enzymes are also added to the food to start breaking down proteins into amino acids.

In the small intestine

Food then passes into the small intestine where more enzymes are added to the food.

The small molecules of food are then absorbed through the walls of the small intestine and into the blood plasma. This process happens by **diffusion**.

f What has to happen to the food so that it can be absorbed through the walls of the small intestine?

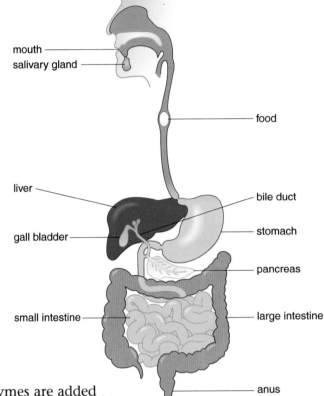

△ The human digestive system

keywords

carbohydrase
• carbohydrate • diffusion •
enzyme • fat • kwashiorkor
• lipase • mineral •
protease • protein • vitamin

Enzyme action

The table below shows which enzymes are added to the food and where.

Food group	Enzymes	Where the enzyme is made	Product
protein	protease	stomach small intestine pancreas	amino acids
fat	lipase	pancreas	fatty acids and glycerol
carbohydrate	amylase	mouth small intestine	simple sugars

g Explain why digestion by enzymes is called chemical digestion.

height: 1.58 m
mass: 100 kg

height: 1.83 m
mass: 59 kg

height: 1.93 m
mass: 77 kg

height: 1.68 m
mass: 64 kg

Health research

Sanjay is doing research for a health food company. He has collected data about different people. Here is some of his data.

Sanjay first has to calculate the body mass index (BMI) of each person. He can do this using this formula:

$$BMI = \text{mass in kg} / (\text{height in m})^2$$

Examiner's tip

Even if you cannot do the maths, make sure you know how to use the BMI chart.

Mass (kg)

Height (cm)	54	59	64	68	73	77	82	86	91	95	100	104	109	113
137	29	31	34	36	39	41	43	46	48	51	53	56	58	60
142	27	29	31	34	36	38	40	43	45	47	49	52	54	56
147	25	27	29	31	34	36	38	40	42	44	46	58	50	52
152	23	25	27	29	31	33	35	37	39	41	43	45	47	49
158	23	24	26	27	29	31	33	35	37	38	40	42	44	46
163	21	22	24	26	28	29	31	33	34	36	38	40	41	43
168	19	21	23	24	26	27	29	31	32	34	36	37	39	40
173	18	20	21	23	24	26	27	29	30	32	34	35	37	38
178	17	19	20	22	23	24	26	27	29	30	32	33	35	36
183	16	18	19	20	22	23	24	26	27	28	30	31	33	34
188	16	17	18	19	21	22	23	24	26	27	28	30	31	32
193	15	16	17	18	20	21	22	23	24	26	27	38	29	30
198	14	15	16	17	19	20	21	22	23	24	25	27	28	29
203	13	14	15	17	18	19	20	21	22	23	24	25	26	28

underweight   healthy weight   over weight   obese

He then has to work out the recommended daily average (RDA) protein intake for each person. This can be calculated using the following formula:

RDA in g = 0.75 × body mass in kg

Because teenagers are still growing, Sanjay knows they should include more protein in their diet than adults who have stopped growing.

Questions

1 Calculate how much protein each of the people should eat each day.

2 Suggest whether your answer should be modified because one of them is a teenager and one is much older.

3 Work out the BMI for each person.

4 Which person is underweight and which person is obese?

Fighting disease

In this item you will find out

- about different things that can make us ill

- how our body responds to illness

- how drugs are tested to make sure they are effective and safe

In the Middle Ages diseases wiped out whole villages – if you look carefully you can still see the remains of the houses

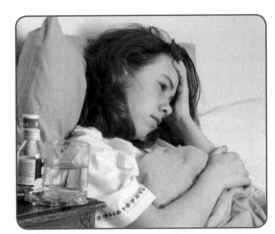

▲ Illness can be caused by many different things

When we are ill we take medicines to make us feel better. Have you ever thought about why we become ill?

Different diseases exist in different parts of the world. This is often due to different climates or whether the country can afford a good health care system and provide clean drinking water and food.

Infectious diseases can be spread by contact with someone who has the disease. Non-infectious diseases can't be spread in this way.

Some diseases are caused by microorganisms. Microorganisms cause disease when they damage the cells in our bodies or produce poisonous chemicals called **toxins**. Microorganisms that do this are called **pathogens**, for example:

- athlete's foot is caused by a fungus
- flu is caused by a virus
- cholera is caused by bacteria
- dysentery and malaria are caused by protozoa.

Doctors can also give patients drugs called **antibiotics**. Antibiotics are effective against diseases caused by bacteria and fungi.

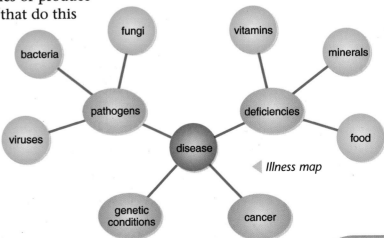

◄ Illness map

a Explain the difference between infectious and non-infectious diseases.

Reducing the risk

Diseases and other disorders can also be caused by a poor diet or problems with your body's make-up:

- scurvy is caused by a vitamin deficiency
- anaemia is caused by a mineral deficiency
- diabetes and cancer are caused by disorders of the body
- red-green colour deficiency is caused by genetic inheritance.

We can reduce the risk of getting some cancers by reducing our exposure to sunlight, stopping smoking and stopping eating certain types of food. Scientists think that eating lots of fresh fruit and vegetables containing anti-oxidants may also reduce our chances of getting cancer. Anti-oxidants work by removing some very dangerous chemicals that are produced by our bodies. The anti-oxidants react with them before they can do any damage to our cells, and to the DNA within cells, which can lead to cancer.

> **Amazing fact**
>
> If you were freeze dried, 10% of the weight would be from microorganisms living on your body.

Malaria

Malaria is one of the world's biggest killer diseases. It is caused by a **parasite** that lives in the blood and liver. The human body and other animals act as a **host** to this parasite. The parasite is spread by a mosquito.

▲ Mosquitoes spread malaria

The mosquito sucks up the blood of an animal that has the parasite. The mosquito then carries the parasite and when it sucks the blood of a human, the parasite passes into the person's body and causes malaria. Animals, such as the mosquito, that carry disease-causing organisms from one animal to another are called **vectors**.

b **Explain how the mosquito spreads malaria.**

Defence against disease

Our bodies have several ways of protecting us from disease:

- skin acts as a barrier to stop pathogens entering our body
- blood clots seal wounds and keep out pathogens
- mucus membranes in our lungs produce mucus to trap dirt and pathogens
- hydrochloric acid in our stomachs kills pathogens in our food
- white blood cells engulf and destroy microbes.

Antigens and antibodies

Antigens are any kind of foreign substance that enters our body. Sometimes they are harmless, such as pollen grains. Sometimes they are harmful, such as bacteria that cause pathogenic diseases. Our bodies respond by making our white blood cells produce **antibodies**. These antibodies lock onto the antigens and stop them from working. If the antigens are pathogens, then the antibodies can even kill them.

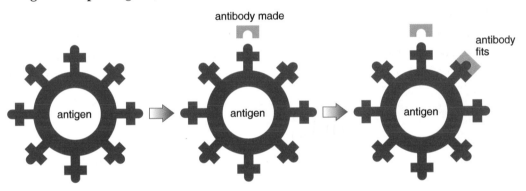

▲ The antigen and antibody reaction

The problem is that it can take several days for our bodies to make the antibodies. During this time we can be very ill or even die. When we have made the antibodies we recover quite quickly. The antibodies remain in our blood so that if the same pathogens invade our bodies again, we will be immune. This is called **active immunity**.

We can also be immunised to give us protection from different pathogens.

Sometimes when we catch a disease our bodies are not capable of making the antibodies in time. This is a life-threatening situation. Fortunately, doctors can sometimes inject us with antibodies (vaccines) that have been made by someone else. This is called **passive immunity**. Unfortunately these antibodies do not last. In order to gain long term protection we need to produce our own antibodies to the disease. Scientists are constantly working to produce new vaccines for diseases such as AIDS and bird flu.

keywords

active immunity • antibiotic • antibody • antigen • host • parasite • passive immunity • pathogen • toxin • vector

c If our bodies are so good at protecting us from disease, explain why people still die from infections.

d Explain the difference between active and passive immunity.

Drug safety

Stephanie is a scientist. She works for a big drug company. Her job is to discover new drugs and treatments to protect us from disease. Stephanie makes sure that all new medical treatments and new drugs are tested before they are used on human patients. This is to make sure that they are safe to use.

Some drugs are first tested on animals. They are then tested on human tissue in the laboratory. Some people object to testing drugs on animals and want to use other methods of testing such as computer models.

One computer model that Stephanie uses is the APACHE simulator. It is a complex computer database that she can use to predict how a drug will work before using it on real people. Some other scientists say that all drugs have side effects and it is much safer to test them on animals first.

This difference of opinion has caused a lot of conflict. Some animals' rights groups believe that it is all right to use violence to protect the laboratory animals from experimentation.

Even when drugs have been passed to use on patients, doctors have a form that they must fill in if a patient reports any adverse reactions to the drug. If there are enough adverse reactions the drug is withdrawn from use.

Questions

1 Suggest why it is important for Stephanie to discover new drugs.

2 Suggest why she must test the drugs before they are given to patients.

3 Explain how drugs can be tested using a computer model.

4 Explain what happens if a patient reports an adverse side effect from a new drug to their doctor.

Messages to the brain

In this item you will find out

- how the brain keeps in touch with all parts of the body

- about how the eye works

- how different parts of the body communicate with each other using neurones

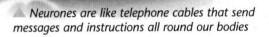

▲ Neurones are like telephone cables that send messages and instructions all round our bodies

Just imagine what it would be like if you lost your senses one by one – to become blind first, then deaf, then to have no sense of feeling and finally no sense of smell or taste. It's hard to imagine what it would be like. In experiments where volunteers are placed in an environment of sensory deprivation, they can usually only last for a few minutes before they press the 'panic button'.

Your body gathers information from the outside world in different ways. You get information from chemicals in food and air, from seeing and touching objects and from hearing sounds.

Your brain receives information about the outside world from your:

- eyes – vision
- skin – touch, pressure, heat and pain
- nose – smell
- tongue – taste
- ears – hearing and balance.

This vast amount of information is then sent along neurones to the brain.

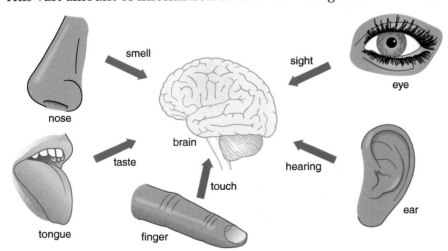

▲ The sense organs inform your brain

Amazing fact

A human brain produces more electrical impulses in a single day than all of the world's telephone networks put together.

▼ *The structure of the human eye*

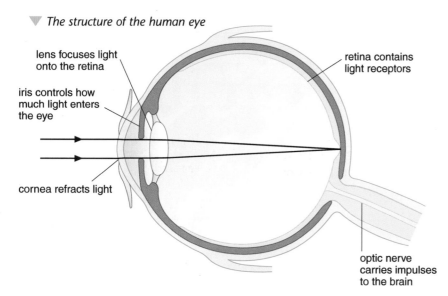

lens focuses light
onto the retina

iris controls how
much light enters
the eye

cornea refracts light

retina contains
light receptors

optic nerve
carries impulses
to the brain

Seeing the light

Your eyes send vast amounts of visual information to your brain.

Eye problems

The retina contains specialised cells that can detect red, green and blue light. These cells enable your brain to see the world in colour. Some people have an inherited condition called colour blindness. This is caused by a lack of these specialised cells in the retina. Often people with this condition can't tell the difference between red and green.

When the eye is working properly, the cornea and the lens refract the light onto the retina at the back of the eye. So you can see things clearly, the lens also focuses the light.

In some eyes the light is focused short of the retina. This is called short sight. In some eyes the light is not yet in focus by the time it reaches the retina. This is called long sight. Long and short sight are caused by the eyeball or the lens being the wrong shape.

 Explain what happens to the rays of light when they are focused by a short sighted eye.

Binocular and monocular vision

Having two eyes on the front of our head gives our brain two separate images of an object. This is called **binocular vision**. It enables us to judge distance quite accurately. Try catching a ball with one eye closed.

We can do this because the brain performs some clever mathematics. As an object approaches us the eyes have to turn inwards. The brain can use this information to judge how far away the object is.

However, having eyes on the front of our head gives us a restricted view of the world around us. We can only see what is in front of us. Some animals have eyes on the sides of their head. They can see all around them but only see things with one eye. This is called **monocular vision**. It lets them see predators coming but is not very good at working out how far away they are.

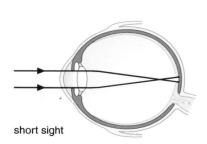

normal eye

short sight

The nervous system

The central nervous system (CNS) consists of the **brain** and **spinal cord**.

The peripheral nervous system (PNS) consists of nerves that carry information into and out of the central nervous system. Nerve impulses travel along nerve cells called **neurones**.

The impulse is electrical and travels along the **axon**.

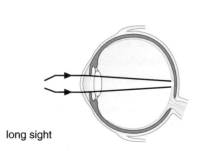

long sight

- **sensory** nerves carry information in the form of electrical impulses from the five senses to the brain
- **motor** nerves carry instructions to our muscles from the brain.

Neurones are the longest cells in our body and can be over one metre in length. The ends of each neurone have many branches. This enables the neurone to connect with many other different neurones producing millions of different nerve pathways.

 b Suggest why neurones can be over one metre in length.

sheath   axon

a motor neurone carries
instructions to our muscles

▲ This neurone
carries instructions
to our muscles

cell body
dendrite

Reflex actions

Reflex actions are very fast actions that our bodies do automatically without having to think about them. They are sometimes used to protect us from danger. This makes them different from voluntary actions that we have to think about and that are under the conscious control of the brain. It takes about a third of a second for an impulse from a sense organ to go to the brain and a return impulse to go to a muscle. If you have just picked up a very hot saucepan, this can be too long. Fortunately this time can be reduced by a **reflex arc**.

In a reflex arc the impulse goes from a **receptor**, along a sensory neurone into the spinal cord. As well as going up to the brain, a **relay** neurone connects directly to the motor neurone. The motor neurone goes to an **effector**, such as the muscle in the arm. This instructs the muscle to let go of the saucepan. By the time the brain receives the pain signal, the hand has already let go of the saucepan. This is called a reflex action.

Other examples of reflex actions include the knee jerk reflex and the pupil reflex. The pupil reflex makes the pupil smaller. It does this by increasing the size of the iris. Once the pupil is smaller it stops too much bright light from entering the eye, which could cause damage.

 c What is the advantage of having a reflex arc?

spinal cord   **Not to scale**

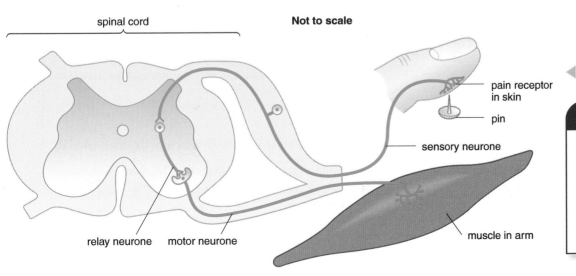

pain receptor
in skin

pin

sensory neurone

◀ The spinal reflex arc

relay neurone   motor neurone

muscle in arm

keywords

axon • binocular vision
• brain • effector •
monocular vision • motor
• neurone • receptor •
reflex arc • relay • sensory
• spinal cord

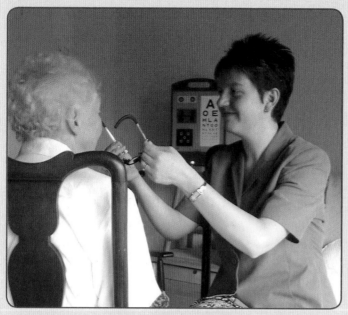

▲ *A visit from the optician*

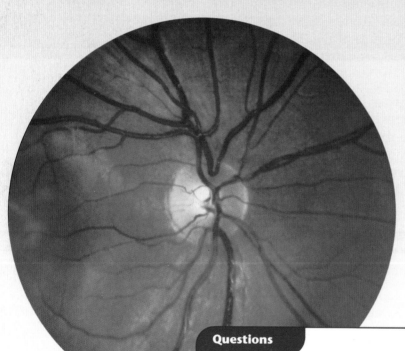

▲ *The retina*

Testing sight

Helen is an optician. Some senior citizens are too old or infirm to go to her, so Helen visits their homes to test their eyes.

It is very important to have eye tests. Most people think that eye tests are performed to check whether you need to wear glasses.

But eye tests can tell the optician a lot about the health of the person. The image of the retina shows what Helen sees when she looks into a healthy person's eye. The retina has a network of fine blood vessels.

Diabetes damages the blood vessels in the retina. It can lead to blindness. Unfortunately in the early stages there are no symptoms. The image of a retina is why it is important that Helen spots the early signs of the disease. She can then refer the person to a doctor who will treat the patient for diabetes.

Helen also tests the pressure inside the eyeball. A blast of air is aimed at the front of the eye and Helen measures how far the cornea is pushed inwards. The higher the pressure in the eye, the less the cornea moves. If the pressure is too high, it is called glaucoma. Like diabetes, it can lead to blindness. This is why it is most important that people have regular eye tests.

Questions

1 Explain why it is important for people to have eye tests.

2 Explain why diabetes and glaucoma can damage the eye before the person realises that there is a problem.

3 Explain how Helen can test for glaucoma.

4 Suggest why, during an eye test, Helen shines a bright light into the person's eye.

5 Suggest how the visual information from a persons retina reaches the brain.

Drugs make changes

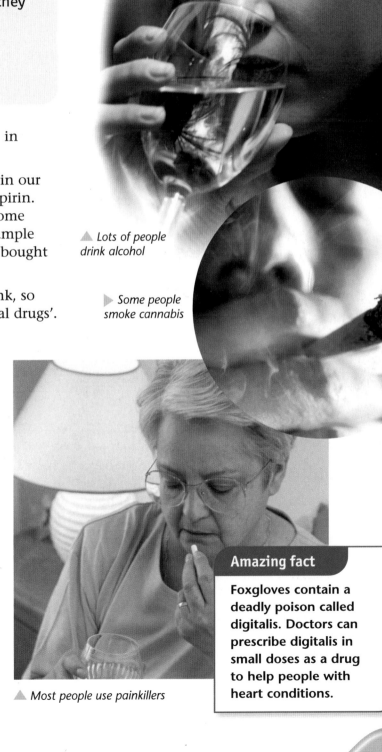

In this item you will find out

- about different types of drugs and how they are classified
- about the effects of different drugs
- about smoking and alcohol

What do alcohol, cannabis and aspirin all have in common? They are all **drugs**.

Drugs are chemicals that produce changes within our bodies. Some drugs can help us, for example aspirin. Some drugs are harmful, for example heroin. Some drugs have to be prescribed by a doctor, for example antibiotics, while others such as aspirin can be bought quite easily.

When people go out they often smoke and drink, so smoking and alcohol are considered to be 'social drugs'.

Drugs such as alcohol can be very **addictive**. As people become more addicted, they also become more **tolerant** to the drug. This means they have to take more of it to get the same effect.

People who are addicted suffer from **withdrawal symptoms** when they stop taking the drug. With some drugs it is sometimes called 'going cold turkey' because just like when you are ill, you go hot and cold and shiver and get goose bumps. It can be a very unpleasant experience.

This is why it is so difficult to stop taking a drug once addicted and why **rehabilitation** can take a long time.

a. Explain what is meant by 'addictive'.

b. Explain what is meant by 'withdrawal symptoms'.

▲ Lots of people drink alcohol

▶ Some people smoke cannabis

▲ Most people use painkillers

Amazing fact

Foxgloves contain a deadly poison called digitalis. Doctors can prescribe digitalis in small doses as a drug to help people with heart conditions.

Types of drugs

Type of drug	What it does	Examples
depressant	slows the brain down	temazepan, alcohol, solvents
stimulant	increases brain activity and helps depression	nicotine, ecstasy, caffeine
pain killer	reduces pain by blocking nerve impulses	aspirin, heroin
performance enhancer	improves athletic performance by increasing muscle development	anabolic steroids
hallucinogen	changes what is seen and heard	LSD, cannabis

Classifying drugs

The law classifies drugs into three different classes.

Class A
These are the most dangerous drugs, such as heroin, cocaine, ecstasy and LSD. Illegal possession of this group carries the heaviest penalties with up to seven years in prison.

Class B
These drugs include amphetamines and barbiturates. Possession of drugs in this group can lead to up to five years in prison. Amphetamines such as 'speed' are stimulants. They work by increasing the activity between different neurones in the brain. Barbiturates are depressants. They slow down the activity between the neurones in the brain.

Class C
These are mainly drugs prescribed by the doctor, and other drugs such as cannabis. They are the least dangerous drugs and carry the lightest penalties if you are caught with them. There is disagreement over drugs such as cannabis. Some people think it should be made legal and freely available. Others think it should be reclassified as a class B drug.

Smoking

Smoking tobacco is highly addictive because it contains a drug called **nicotine**. Addiction to the drug makes it hard to stop once you start. When you burn a cigarette, nicotine, carbon monoxide, tars and tiny particles called particulates are produced.

If you smoke you run the risk of developing these diseases:

- emphysema
- bronchitis
- cancer of the lung, throat, mouth or oesophagus
- heart disease.

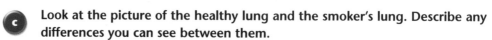

c **Look at the picture of the healthy lung and the smoker's lung. Describe any differences you can see between them.**

Tar, particulates or tiny particles and carbon monoxide all reduce the amount of oxygen that our bodies absorb from the air. This can lead to shortness of breath and heart disease.

▲ Healthy lung (below) and smoker's lung (above)

Smoker's cough and bronchitis

The trachea, bronchi and bronchioles in the lungs are lined with mucus to trap dirt and microbes, and small hairs called **cilia** found on epithelial cells. The job of the cilia is to waft mucus up from the lungs to the back of the throat where it is swallowed. Cigarette smoke contains chemicals that stop the cilia from working. This means that mucus accumulates in the lungs. The only way to get rid of it is to cough. This is called a 'smoker's cough'. The build up of mucus in the bronchi can become infected and cause bronchitis.

d What damage does smoking do to our lungs?

e Look at the graphs.

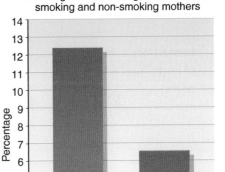

Amazing fact

90% of people who die from lung cancer are smokers.

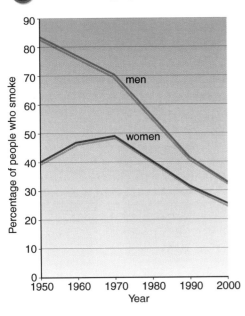

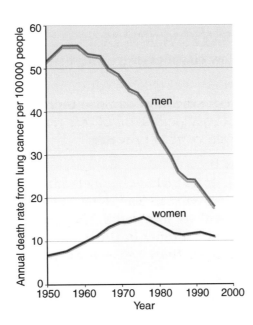

Percentage low birth weight in babies from smoking and non-smoking mothers

(i) Describe what effect smoking has on the birth weight of babies.
(ii) Who smokes the most cigarettes, men or women?
(iii) Explain what happened to the number of deaths from lung cancer when the percentage of people who smoked began to fall.

Drinking alcohol

Alcohol is a poisonous drug that is removed from the body by the liver. It has short-term effects on the brain and nervous system:

- it impairs our judgement
- it upsets our balance and muscle control
- it blurs our vision and speech
- it causes blood to flow in the surface of the skin leading to heat loss
- it makes us sleepy.

f Use the information given above to explain why there is a legal limit for the level of alcohol in the blood, for car drivers and pilots.

Long-term use can damage the liver causing a disease called **cirrhosis** of the liver. It can also damage the brain and nervous system. As the liver is responsible for a large number of the functions that take place in the body, liver damage is very serious and can lead to death.

keywords

drug • addictive • tolerant
• withdrawal symptoms
• rehabilitation • nicotine
• cilia • cirrhosis

Drink-driving

Tina, Su and Angus are out at a party one weekend. It is getting late and Tina wants to go home. Angus says he has his car and he will drive her home. They are just about to leave, when Su stops them. She says that Angus has had too many alcoholic drinks and that he is over the limit for driving. Angus says that he has only had a couple of drinks and that he will be fine.

Su knows that alcohol consumption is measured in units. The drinks in the photograph all contain one unit of alcohol each.

▲ Each drink contains one unit of alcohol

beer
1 pint

2 units

beer
½ pint

1 units

wine

1 unit

spirit

1 unit

▲ What Angus has had to drink

If Angus has had four or more units of alcohol then he could be charged with drink-driving if he is stopped by the police. Angus says he has had the following drinks: half a pint of beer, a glass of wine, another half pint of beer, a glass of vodka and a final half pint of beer.

Questions

1 Work out how many units of alcohol Angus has had to drink.

2 Is Angus over the drink-driving limit?

3 Do you think that people should be allowed to drink any alcohol at all before they drive? Explain your answer.

Bodies don't like change

In this item you will find out

- why keeping a constant internal environment in our bodies is important

- how the body keeps a constant temperature

- how hormones are used to control how our body works

Humans live in environments that are constantly changing. Sometimes it is hot, sometimes cold. Sometimes lots of food is available, sometimes it is not. In order for our bodies to work properly they need to maintain a constant internal environment. This is called **homeostasis** and it means balancing what we take into our bodies with what we give out.

Our bodies work hard to maintain constant internal levels of water, oxygen and carbon dioxide. They also work hard to maintain a constant temperature. These factors are controlled by automatic control systems so that our cells can function at their optimum level.

The reason why humans are so successful and have colonised every part of the planet is because of homeostasis. Most animals or plants are only found in specific areas. Polar bears are found at the North Pole and cacti are found in the desert. But humans are found everywhere.

Some animals that cannot control their own internal environment are often only found in parts of the world where conditions are just right. Others just shut down and go dormant during times when conditions are not to their liking. Even animals that can control their internal environment, like the polar bear, are so adapted to their external environment that they can only be found in certain places in the world. All that fur would make it very hot for the polar bear if it lived at the equator. Unlike the polar bear, we can take our warm furry coats off.

▲ *Clothes that protect against the Sun*

▲ *Clothes that protect against the cold*

Amazing fact

Your body will respond to a change in air temperature of less than 1 °C.

a Explain why, unlike some other animals, humans can live in both hot and cold countries.

Controlling body temperature

A healthy person has a constant body temperature of about 37 °C. The skin is the organ that is responsible for controlling this temperature.

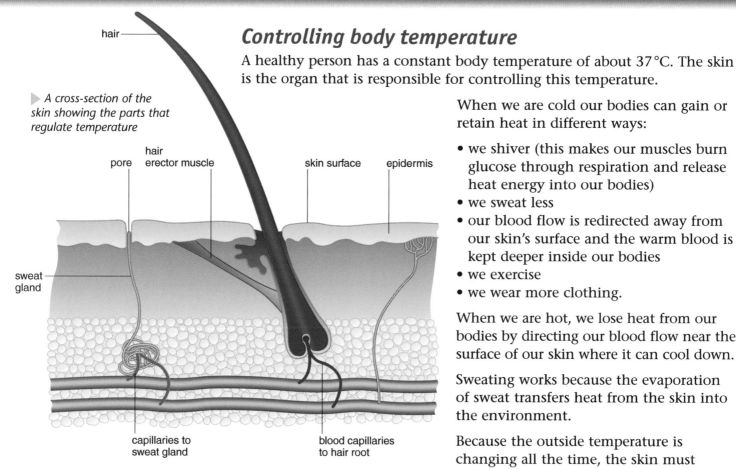

▶ *A cross-section of the skin showing the parts that regulate temperature*

hair

pore

hair erector muscle

skin surface

epidermis

sweat gland

capillaries to sweat gland

blood capillaries to hair root

When we are cold our bodies can gain or retain heat in different ways:

- we shiver (this makes our muscles burn glucose through respiration and release heat energy into our bodies)
- we sweat less
- our blood flow is redirected away from our skin's surface and the warm blood is kept deeper inside our bodies
- we exercise
- we wear more clothing.

When we are hot, we lose heat from our bodies by directing our blood flow near the surface of our skin where it can cool down.

Sweating works because the evaporation of sweat transfers heat from the skin into the environment.

Because the outside temperature is changing all the time, the skin must maintain a delicate balancing act between heat lost and heat kept inside.

Extreme temperatures

If you are somewhere which is very hot or very cold it might become difficult for you to control your body temperature and this can be dangerous.

If your body temperature gets too high, you can suffer from **heat stroke**. This can lead to **dehydration**. If this is left untreated you could die.

If your body temperature gets too low you can suffer from **hypothermia**. This also can cause death if it is not treated.

Amazing fact

Every square inch of skin contains 20 feet of blood vessels that help to control our temperature.

Hormones

Hormones are chemicals in our bodies that transmit instructions from one part to another. They travel in the blood to their target organs.

One of their uses is to maintain homeostasis within the body. Because hormones travel in the blood they take longer to have an effect than nervous reactions.

Hormones are produced by the human **endocrine** glands.

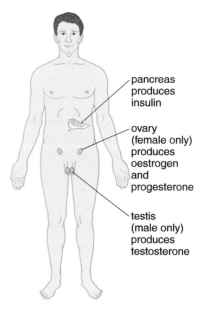

pancreas produces insulin

ovary (female only) produces oestrogen and progesterone

testis (male only) produces testosterone

▲ *Where some hormones are produced*

Sex hormones

The male and female sex hormones (testosterone, oestrogen and progesterone) are responsible for the secondary sexual characteristics that happen at puberty.

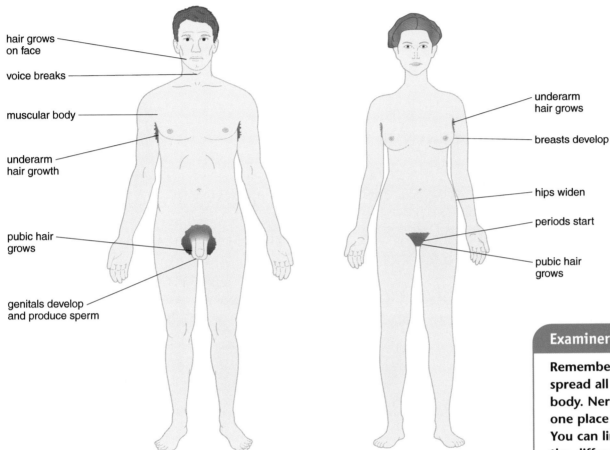

hair grows on face

voice breaks

muscular body

underarm hair growth

pubic hair grows

genitals develop and produce sperm

underarm hair grows

breasts develop

hips widen

periods start

pubic hair grows

▲ The effects of male and female sex hormones

Human **fertility** can be controlled by the artificial use of sex hormones. **Contraceptive pills** and fertility drugs all contain sex hormones.

Diabetes

Some people suffer from **diabetcs**. They do not produce enough of the hormone insulin in their pancreas.

Insulin controls the blood sugar levels. People who do not produce enough insulin can have too high a blood sugar level, which can be dangerous.

Diabetics have to be careful that they do not eat too much sweet food. They may also need to inject themselves with the hormone insulin to help them control their blood sugar levels.

 b **Explain the job of the hormone insulin.**

keywords

contraceptive pill • dehydration • diabetes • endocrine • fertility • heat stroke • homeostasis • hormone • hypothermia • insulin

Measuring our temperature

Scientists have developed different devices for measuring our temperature.

▲ Clinical thermometer

▲ Digital thermometer with temperature probe

	°C
blue	34
green	35
yellow	36
red	37

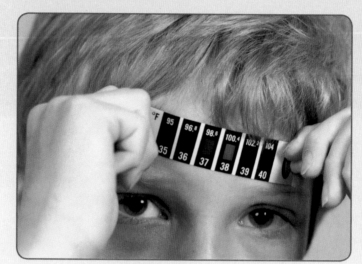

▲ Temperature colour strip

▲ Thermal image

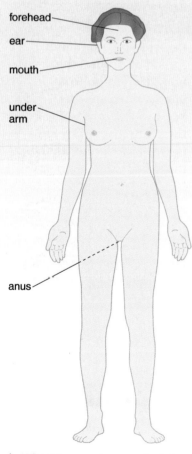

forehead
ear
mouth
under arm
anus

▲ Where temperature can be taken

Our temperature can be also measured in different parts of our bodies. Although our average body temperature is 37 °C, it varies slightly in different parts.

Questions

1 One of the devices for measuring body temperature shows that the person is ill. Which one is it?

2 Suggest one advantage of using a colour-sensitive strip with a small child rather than using a clinical thermometer.

3 Suggest why a clinical thermometer may be more accurate than a colour-sensitive strip.

4 Which of the devices in the above pictures does not need to touch the person to get a temperature reading?

5 Suggest why constant temperature monitoring of someone who is unconscious is easier with a digital meter and temperature probe.

It's all in the genes

In this item you will find out

- about chromosomes and genes
- about DNA
- about sexual and asexual reproduction

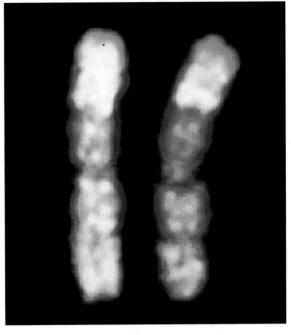

There have been lots of stories in the news over the past couple of years about genetics and designer babies. People argue about whether scientists should be allowed to alter human genes.

It is easier to make up your mind about the pros and cons of genetic engineering if you understand the science behind some of the stories.

The nucleus of a body cell contains **chromosomes**. They carry information in the form of **genes**. This information in the genes is held as a set of coded instructions called the **genetic code**.

Your genetic code controls the activity of your body cells. This means that your genetic code also controls some of your characteristics.

Genes are made from a chemical called **DNA**. Chromosomes are whole strings of genes placed side by side.

The chromosomes contain all the information needed for making a new human being.

DNA is an amazing chemical. It must code for all the information needed to make a new human being. It must be able to copy itself so that the information can be passed on to future generations. And it has to be small enough to be stored inside the nucleus of nearly every cell in our body.

In order to fit all the DNA into each nucleus, the DNA is coiled many times to make it shorter. Just as the filament in a light bulb can be over a foot in length, coiling it up makes it shorter. The difference with DNA is that each coiled is coiled again and again to make it very short indeed.

a Explain how such a long molecule as DNA can be squeezed into the nucleus of all our cells.

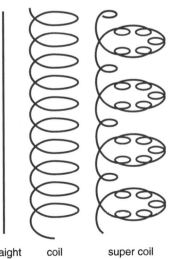

straight   coil   super coil

▲ The coiling in DNA

Amazing fact

Each human body contains enough DNA to stretch to the Moon.

▲ Baby Rowan

DNA in detail

In order to code for all the information, DNA uses four different chemicals, or **bases**. These four chemicals are called A, T, C and G.

It is the order of these bases that stores the code or instructions for making a new human being. Each gene contains a different sequence of bases and codes for a single instruction.

Unlike the English language that has 26 letters in its alphabet, DNA only has four bases in its alphabet. This means that, unlike words that are made up of about five or six letters, genes are made up of hundreds of bases. This makes chromosomes very long indeed.

In this four-letter code for the gene that makes the hormone insulin, each base is represented by a letter (a,t,g,c):

atggccctgtggatgcgcctcctgcccctgctggcgctgctggccctctggggacctgacccagccgcagcctt tgtgaaccaacacctgtgcggctcacacctggtggaagctctctacctagtgtgcggggaacgaggcttcttct acacacccaagacccgccgggaggcagaggacctgcaggtggggcaggtggagctgggcggggccctgg tgcaggcagcctgcagcccttggccctggaggggtccctgcagaagcgtggcattgtggaacaatgctgtacc agcatctgctccctctaccagctggagaactactgcaactag

Just imagine how long the sequence of bases would need to be to make baby Rowan, who has over 30,000 genes, many of which are much longer than the one for insulin.

b Explain why, unlike words in English, the instructions needed to code for a gene are so long.

Sexual reproduction

Just like his parents, Rowan has 23 matching pairs of chromosomes in the nucleus of each cell. This is because his mother's egg and his father's sperm (called **gametes**) both contained 23 single chromosomes. When the gametes joined together the full set of 23 pairs was restored. This is called sexual reproduction.

23   23

fertilisation

46

first cell of new baby

▲ A mother's egg and father's sperm

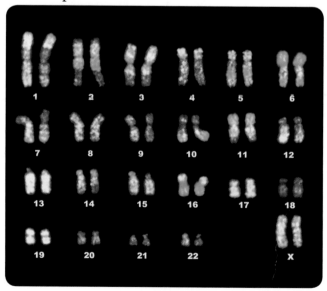

1	2	3	4	5	6
7	8	9	10	11	12
13	14	15	16	17	18
19	20	21	22		X

▲ Humans have 23 pairs of chromosomes

▲ These people have different genes

▲ These twins have the same genes

Different species have different numbers of chromosomes but they are always an even number.

c **Suggest why the number of chromosomes is always an even number.**

The world is full of variation. Apart from identical twins who have the same DNA, no two humans look the same. This is because sexual reproduction always produces variation.

The twins in the photograph are identical because after **sexual reproduction** had taken place the **fertilised** cell split into two separate cells. Each cell then went on to divide to produce two identical humans.

d **What is unusual about the genes in identical twins?**

▲ This animal grows identical babies from its own body

Identical offspring

Some organisms reproduce using asexual reproduction. This produces organisms that have the same genes as their parents. They look exactly like their parents. Look at the photographs. They show some examples of **asexual reproduction**.

Animals and plants that produce identical copies of themselves are said to produce **clones**.

e **List three examples of animals and plants that reproduce by asexual reproduction.**

f **What are identical copies of organisms called?**

▲ This plant also grows identical copies

▲ Greenfly can produce thousands of copies of themselves without sexual reproduction

keywords

asexual reproduction
• base • chromosome
• clone • DNA • fertilise
• gamete • gene • genetic
code • sexual reproduction

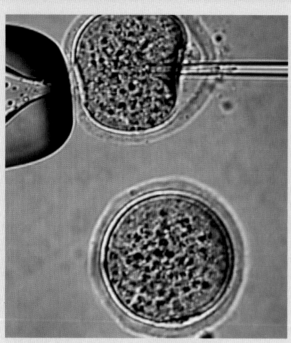

▲ *Selecting a cell from an embryo*

Designer babies

A designer baby is created by removing a single cell from an embryo in the first five days after fertilisation. It is then genetically tested before deciding whether to implant it into the mother's womb.

By doing this it is possible to:

- screen for diseases such as cystic fibrosis and Down's syndrome
- create a healthy baby to treat a sick brother or sister
- select the sex of the child.

A British couple tried to get permission to genetically select an embryo which would be a near perfect genetic match to their four-year-old son, who had a life-threatening blood disorder. When the child was born, cells from the umbilical cord could be used to cure their four-year-old son.

Permission was refused in this country, so the couple went to the USA to have the procedure carried out.

It wasn't until 2005 that scientists were given permission to carry out the procedure in this country.

It sounds a brilliant way of saving a child's life, but not everyone agrees.

Arguments against:

1. it is unethical and may lead to human cloning
2. it treats the baby as a thing and not a person
3. it is not right to subject the newborn baby to medical procedures for someone else's benefit
4. embryos that are not suitable will be discarded – some people think this is murder.

Arguments for:

1. it saves lives and prevents suffering
2. if the new baby is wanted and loved, it is ethically OK
3. the new baby is not injured, only cells that would be thrown away with the umbilical cord are used
4. abortion is legal up to 24 weeks – the embryos that are discarded are only a ball of cells a few days old.

Examiner's tip

With questions on this subject you should always give both sides of the argument in your answer.

Questions

1 Write down two arguments against having designer babies.

2 Write down two reason for having designer babies.

3 Do you think designer babies are a good idea? Explain your reasons.

Uniquely you

In this item you will find out

- which human characteristics are controlled by genes and which are not

- about genetic variation

- about mutations

Scientists are now trying to decide the relative importance of our genes, as opposed to the environment, in making us who we are.

We know that it is usually a combination of both factors, but not how much each contributes. If we get the right genes we may be good at sport, good at school or very healthy.

But we also know that to be good at sport requires hours of training, to be good at exams requires a lot of revision and to be healthy requires us to eat the right foods and not to smoke.

Speaking French involves inheriting the ability to speak and understand a language, but which language we learn depends on where we are born.

Some characteristics are controlled only by the environment and not by our genes. These include scars, tattoos and having a sun tan.

Other characteristics are only controlled by our genes. We inherit nose shape, the size of our earlobes and the colour of our eyes from our parents.

 Which of the following characteristics are only controlled by the environment: intelligence, height, scars, nose shape, weight and sun tan?

I can't help being outrageous – it's in my genes!

Amazing fact

Apart from identical twins, no two people who have ever been born on Earth look exactly the same.

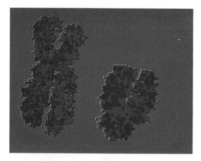

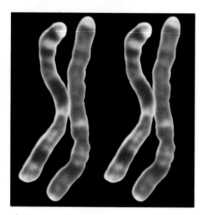

▲ Male chromosomes

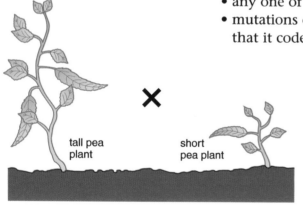

▲ Female chromosomes

Male or female

Sex inheritance is controlled by a whole chromosome, rather than a single gene.

In humans, males have two different chromosomes called X and Y. The X chromosome is larger than the Y chromosome.

Females have two chromosomes that are the same called X and X.

We can use a punnet square to see how sex is inherited.

		mum	
		X	X
dad	X	X X	X X
	Y	X Y	X Y

b What is the ratio of boys to girls born?

Genetic variation

Genetic variation can occur in three ways:

- when gametes are made they all contain a different combination of genes
- any one of millions of male sperm can fertilise the female egg
- mutations can change the structure of the DNA and alter the instructions that it codes for.

tall pea plant

× short pea plant

offspring all tall

Inherited disorders

Some conditions are caused by having faulty genes. Just like the good genes, the faulty genes can be inherited as well. Conditions that are caused by faulty genes include red-green colour blindness, sickle-cell anaemia and cystic fibrosis.

Dominant or recessive?

We know from the previous topic that our cells contain two complete sets of instructions, one from mum and one from dad. Some of these instructions are **dominant**. Others are **recessive**.

We can usually identify which characteristics are dominant and which are recessive by looking at the number of different types produced.

For example, when two true-breeding tall pea plants are crossed with short pea plants, all the offspring are tall. This tells us that tall is dominant. And short is recessive.

c In pea plants is tall or short the recessive gene?

More about mutations

DNA is a very delicate chemical and can easily be damaged. Changes to DNA that damage it are called **mutations**. Mutations occur when some of the bases are removed, or even moved to a different position in the gene.

Mutations are nearly always bad news as the message in the DNA becomes disrupted.

Imagine you went through this textbook and took out some of the letters, or replaced some other letters with different ones. The book would very quickly become unreadable and useless.

It is the same when mutations happen to DNA. The mutation prevents the gene from producing the protein that it normally codes for.

On very rare occasions however, the mutations can be useful. By pure chance a change may not produce gibberish but may alter the message to make a different one. For example:

Urgent message. Send more guns and ammunition.
Urgent message. Send more buns and ammunition.

Sometimes this can be very useful as it can produce even more variation within a species.

Mutations can be caused by many different things:

• ultraviolet light in sunshine or sun beds
• chemicals in cigarette smoke
• chemicals in the environment
• background radiation in the environment.

All of these factors can damage and change the sequence of bases in our DNA.

Because we have so much spare unused DNA the chances are that we will not notice most of these mutations. However some mutations can cause diseases such as cancer.

 List four causes of mutations.

▲ *The stripy colours in this rose are caused by a mutated gene*

keywords

dominant • mutation
• recessive

Genetics of the fruit fly

Jake is a geneticist. He discovers how genes work by studying the fruit fly.

Jake knows that some fruit flies have abnormally small wings and cannot fly properly.

He crosses a fruit fly with normal wings with a fruit fly with short wings.

Jake discovers that all the offspring have normal wings.

▲ Fruit fly

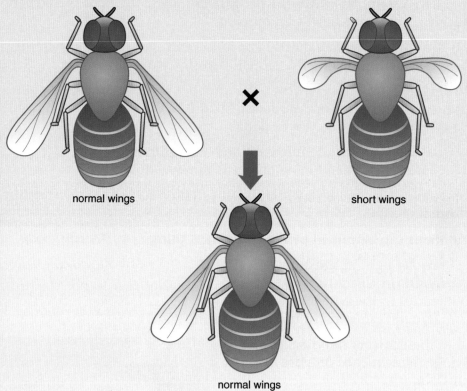

normal wings   ×   short wings

normal wings

Jake decides to use a capital W to represent normal wings.

He uses a small w to represent short wings.

He crosses two of the offspring together.

Jake draws a punnet square to show the result of his experiment.

	male	
	W	w
female W	WW	Ww
female w	Ww	ww

Questions

1 Which characteristic, normal or short wing, is dominant? Explain your answer.

2 What proportion of the offspring from the second cross have short wings?

3 Suggest what would happen if Jake crossed two flies together that both had short wings.

B1a

1 The blood in our body is under pressure.

  a Which of the following best explains the cause of this pressure?
    A gravity, pulling the blood downwards
    B muscles in the skin squeezing the blood vessels
    C contractions of the heart
    D drinking too much water   [1]

  b Suggest why the blood is under pressure.   [1]

2 Finish the sentences by using words from the list.

age   diastolic   mm Hg   systolic

Blood pressure is measured using the units ____(1).
Blood pressure varies according to ____(2).
Blood pressure when the heart is contracting is called ____(3) pressure.   [3]

3 Fitness and health are not the same. Which of the following descriptions best describes fitness and which describes health?

    A not having any form of disease or illness
    B feeling good about yourself
    C taking regular exercise
    D feeling wide awake   [2]

4 Which of the following word equations best describes what happens during hard exercise?

    A glucose + oxygen → energy
    B glucose → lactic acid + carbon dioxide + energy
    C glucose → lactic acid + energy
    D glucose → water + lactic acid + energy   [1]

B1b

1 Finish the sentences by using the words from the list.

**balanced   constipation   energy   growth
haemoglobin   repair   scurvy**

A diet that contains all the food types and nutrients that we need is called ____(1).

Carbohydrates and fats are used to provide ____(2).
Proteins are used for ____(3) and ____(4).
Vitamin C prevents ____(5).
Fibre prevents ____(6).   [7]

2 Finish the sentences by using words from the list.

amino acids   fatty acids   glycerol   simple sugars

Carbohydrates are made up of ____(1).
Fats are made up of ____(2) and ____(3).
Proteins are made up of ____(4).   [4]

3 The following statements about digestion are in the wrong order. Write them out in the correct order.

    A enzymes break food down in the small intestine
    B food is chewed
    C small molecules are absorbed through the gut wall by diffusion
    D food is mixed and churned in the stomach   [4]

4 Look back at the formula for calculating the recommended daily allowance of protein on page 10.

  a Explain how the recommended daily allowance of protein for a person is calculated.   [1]
  b Calculate the RDA of a person with a body mass of 80 kg.   [1]
  c Suggest how the RDA may vary slightly with age. [1]

B1c

1 Look at the diagram of the human body.

Explain how each of the labelled areas helps to defend us against pathogens.

mucus membrane
stomach
blood
skin

    [4]

2 State which of the following parts of the blood help to protect us from disease.

    A plasma   B platelets
    C red blood cells   D white blood cells   [1]

3 Mosquitoes spread malaria.

  a What are organisms called that spread disease from one organism to another?   [1]
  b The following statements are in the wrong order. Write them out in the correct order.

    A The malarial parasite develops in the mosquito.
    B The person develops malaria.
    C A harmless mosquito sucks blood from an animal with malaria.
    D A mosquito carrying the parasite sucks blood from a healthy human.   [4]

4 There are many different kinds of disease.

For each of the following examples, name a specific disease.

A vitamin deficiency   B body disorder
C genetic   [3]

5 Finish the sentences by using words from the list.

active antibiotics antibodies antigens damage passive toxins

Pathogens produce ____(1) and can cause cell ____(2).
Pathogens have ____(3) which are locked onto by ____(4).
After we recover from an infectious disease we often have ____(5) immunity to that disease.
Bacterial and fungal infections can also be treated using ____(6).   [6]

B1d

1 Copy the diagram of the eye and complete the missing labels.   [5]

2 Finish the sentences by using words from the list

brain electrical neurone reflex spinal cord

The impulse in nerve cells is mainly ____(1).
The CNS consists of the ____(2) and ____(3).
The correct name for a nerve cell is a ____(4).
The knee jerk is an example of a ____(5) action.   [5]

3 Explain the function of each of the following parts of the eye.

A cornea   B iris   C lens
D retina   E optic nerve   [5]

4 Look at the diagram of the reflex arc.

spinal cord   (Not to scale)   pain receptor in skin
pin
sensory neurone
relay neurone   motor neurone   muscle in arm

List an example of each of the following from the diagram.

A stimulus   B sensor   C effector   [3]

5 Neurones are highly adapted to the job that they do. List one way in which a neurone is adapted to its job.   [1]

B1e

1 For each of the following types of drug, describe its effect on the human body.

A depressant   B pain killer
C stimulant   D performance enhancer   [4]

2 State which of the following diseases can be caused by smoking.

A cancer   B emphysema   C colour blindness
D bronchitis   E malaria   [3]

3 State which of the following are **not** caused by alcohol.

A blurred vision   B slurred speech
C impaired judgment   D faster reaction times
E damaged liver   [1]

4 Finish the sentences by using words from the list.

bronchi bronchitis cilia cough mucus

Cigarette smoke stops small hairs called ____(1) from working.
The hairs normally propel sticky ____(2) up from the ____(3) to the back of the throat. A build up of sticky substance in the lungs can cause ____(4) and a smoker's ____(5).   [5]

5 Alcohol consumption is measured in units.

beer 1 pint
wine
spirit

a How many units are there in three pints of beer?   [1]
b If a person drinks three pints of beer a night, how many units will they consume in one week?   [1]
c Men should not consume more than 21 units each week. If a man drinks 3 glasses of wine and one glass of whisky each day, will they be consuming too much alcohol?   [1]
d Women should not consume more than 14 units each week. Suggest why the unit limit for woman is less than the unit limit for men.   [1]

B1f

1 Homeostasis is when the body maintains a constant internal environment.

a List three things that the human body works to keep constant.   [3]
b Which of the following is normal body temperature?

A 27°C   B 30°C   C 37°C   D 40°C   [1]

2 a Describe three different ways of monitoring body temperature.   [3]

b Explain the advantages and disadvantages of each method.   [3]

3 Hormones are chemicals produced by the body.

   a How do hormones travel round the body? [1]

   b State one difference between hormones and neurones. [1]

   c Look at the drawing of the human body. State which of the labels, A, B or C, is the pancreas, ovaries and testes. [1]

4 Finish the sentences using the following words.

close evaporates heat sweating open

When the body is too hot, ____(1) occurs which ____(2) and causes the body to cool. Blood vessels near the surface of the skin ____(3) causing the skin to go red and radiate ____(4) away. [4]

5 Oestrogen and testosterone are sometimes called the secondary sexual hormones.

   a Explain what this means. [1]

   b describe the effects on a young teenager of:
     (i) oestrogen   (ii) testosterone [2]

B1g

1 Our genetic code is found in the nucleus of almost every cell of our bodies

   a Put the following structures in their correct order of size, starting with the smallest first.

     nucleus gene DNA molecule
     chromosome [4]

   b Which of the following statements are true?

     A Genes are made of a chemical called DNA.
     B Many chromosomes are joined together to form a gene.
     C A nucleus is found inside every chromosome.
     D Most cells contain chromosomes in matching pairs. [1]

2 In a new baby, what proportion of genes comes from each parent?

   A all from mum
   B all from dad
   C half from mum and half from dad
   D none from mum or dad [1]

3 Chromosomes are found inside the nucleus.

   a How many chromosomes are found in the nucleus of most human cells? [1]

   b Is this number the same for all living organisms? [1]

   c What is unusual about the number of chromosomes found in all living organisms? [1]

4 The information required to make a human being is coded in DNA.

   a How many letters are in the DNA alphabet? [1]

   b Explain how a complete set of DNA manages to fit inside the nucleus of a cell. [1]

5 Which of the following statements about gametes is true?

   A Gametes contain the same number of chromosomes as other body cells.
   B Gametes contain twice the number of chromosomes as other body cells.
   C Gametes contain half the number of chromosomes as other body cells.
   D Gametes do not contain any chromosomes. [1]

B1h

1 State which of the following characteristics are controlled by our genes.

height eye colour scars nose shape tattoos [2]

2 List three characteristics that are controlled by both our genes and our environment. [3]

3 Some disorders can be inherited if a gene becomes changed or damaged.

   a List three disorders that can be inherited. [3]

   b What do we call a gene that has been changed? [1]

4 Look at the following diagram. It shows how sex is determined.

	X	X
X	XX	XX
Y	XY	XY

   a Which are the sex chromosomes that determine males? [1]

   b Which are the sex chromosomes that determine females? [1]

   c Use the diagram to explain why equal numbers of boys and girls are born. [1]

5 Two people who could roll their tongue married and had children. Three of their children could roll their tongue and one could not.

State which condition is dominant and which is recessive. Explain your answer. [3]

6 Mutations are changes to the DNA in genes.

   a Which of the following can cause mutations to DNA?

   water radiation chemicals sound [2]

   b Explain whether most mutations are harmful [1]
     or beneficial. [3]

B2 Understanding our environment

What's the point of looking after the Earth if sometime soon an asteroid is going to crash into it and destroy all living things?

Don't be daft. It might be millions of years before that happens.

Yes it might. Anyway humans are doing more damage to the planet right now and that is much more important to worry about.

- This module is about understanding our environment so that we can all live as part of it and maintain it for future generations. It is only by understanding our environment that we can preserve and protect it.

- In this module you will learn how different animals and plants are adapted and compete with each other so that their populations are maintained. You will also learn how organisms evolve to fit a new and changing environment.

- Sadly, humans are now having a major impact on the environment, making changes that will soon be too late to reverse. It is important to know what these environmental changes are and the decisions that society has to make in order to ensure that we protect and conserve our very delicate world.

What you need to know

- All living things are different and can be put into groups.

- Environmental, ecological and feeding relationships exist between different species of plant and animal.

- Plants are the source of all food and make it by the process of photosynthesis.

Pieces in a jigsaw

In this item you will find out

- about different ecosystems

- how data about ecosystems can be collected

- how to use keys to identify different animals and plants

An **ecosystem** is a place or **habitat**, together with all the animals and plants that live there. The individual plants and animals form the **community**. A breeding group of animals or plants in the ecosystem is called a **population**. There are many different kinds of ecosystem.

Some are natural.

Some are artificial.

We know more about the surface of the Moon than we know about some of the Earth's ecosystems, such as the ocean depths. There are possibly many undiscovered **species** down there.

Natural ecosystems

Amazing fact

There are more different kinds of insect on Earth than all the other different kinds of living things put together.

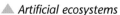
Artificial ecosystems

Bill didn't really understand what kind of key to use

Identifying organisms

One way to identify all the different animals and plants in an ecosystem would be to compare the living **organisms** with lots of different photographs of animals and plants to find a match. This could take a long time. There is a large variety of plants and animals even in a small area such as a square metre.

a How long would it take to identify an insect if you looked at one million photographs and it took just one second to look at each one?

A much faster way is to use a **key**. Keys work by grouping organisms. Each group is then divided into smaller groups. This may sound complicated, but usually it only takes a few divisions to be able to identify an organism.

Try using this key to identify the following animals.

Key		
1	has legs	go to 2
	has no legs	slug
2	has six legs	go to 3
	has eight legs	spider
3	all legs nearly the same length	go to 4
	back legs much longer	go to 5
4	wings visible	damselfly
	no wings visible	beetle
5	shorter antenna	cricket
	longer antenna	grasshopper

Examiner's tip

Practise using keys until you find them easy.

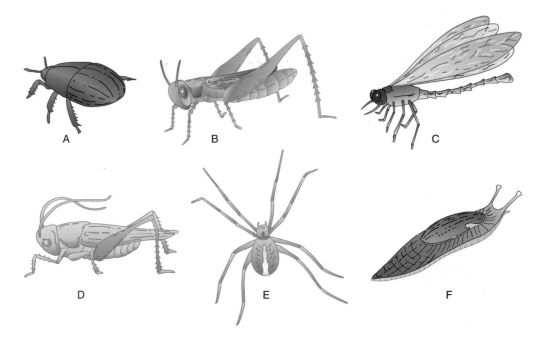

A   B   C

D   E   F

b Imagine you were making a key to identify a new member of your class. What are the first two groups that you could divide the class into?

When you are identifying organisms using a key, always check your answer. Organisms are usually found only in specific habitats. It would be a surprise if you identified a tiger living in your school grounds.

Data collection

One way of collecting data, for example on a school playing field, is to use a **quadrat**. A quadrat is a metal or plastic square that encloses 0.25 square metres. The quadrat is thrown at random and all the different species found within it are identified and counted.

This is much easier than counting organisms for a whole playing field. You can then multiply the numbers counted by four to get the answer in square metres, and then multiply the answer by the number of square metres in the playing field.

c If a student counts three beetles inside one quadrat and the playing field is 10 000 square metres, how many beetles would you expect to find in the playing field?

Other ways of collecting organisms include **dip nets** for ponds and rivers.

▲ A student using a quadrat

▲ A student using a dip net

Pooters can be used to collect small insects.

▲ A student using a pooter

Pitfall traps can be used to catch larger insects.

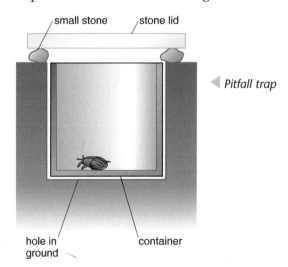

small stone   stone lid

◀ Pitfall trap

hole in ground   container

The changing environment

Jess and Sanjay are doing a research project about the wildlife found around Gateway Valley School.

They have found two maps which show the environment around the school. The left-hand map is about 30 years old. The right-hand map is up to date.

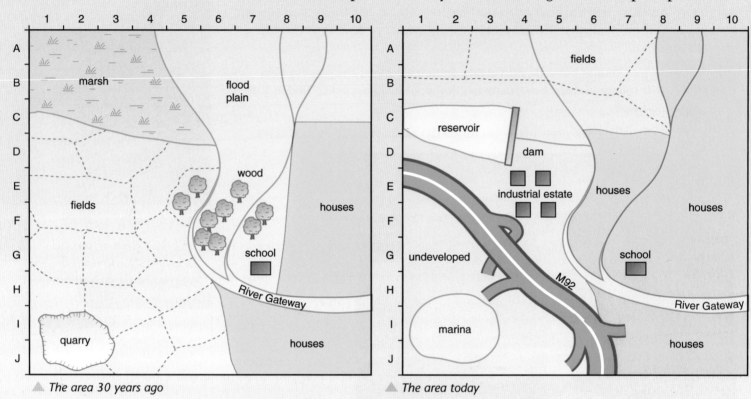

▲ *The area 30 years ago* ▲ *The area today*

They have asked their local wildlife club to give them data about which species were found 30 years ago and which species are found now. The wildlife club has emailed them the following data.

Questions

1 Why do you think badgers are no longer found in area F6?

2 What changes have happened to area I2?

3 Suggest why field mice have not completely disappeared from area E1.

4 Kestrels eat mice. Suggest why the number of kestrels has increased near the motorway.

5 Suggest why building a wildlife pond in the school grounds would increase the number of species present in the area.

Map ref	Species	30 years ago	Today
B2	bull rush	lots	none
	wheat	none	lots
	marsh warbler	few	none
	moorhen	few	none
F6	bluebell	lots	few
	owl	few	none
	sparrow	few	lots
	rose	none	few
	badger	few	none
E1	wheat	lots	none
	barley	lots	none
	field mouse	lots	few
	kestrel	none	few
A9	meadow blue	lots	none
	bee	lots	few
	orchid	few	none
	buttercup	lots	few

Pigeon-holing organisms

In this item you will find out

- how living organisms are classified into different groups

- what is meant by the word species

There are more **species** alive today than at any other time in the history of our planet. New species are being discovered all the time.

There are so many new species waiting to be discovered that it is possible that one of you reading this page will discover a new species and have it named after you.

A scientist called Andrew Polaszek has just discovered a new species of whitefly. The whitefly looks like it is smiling. It is called a smiling whitefly.

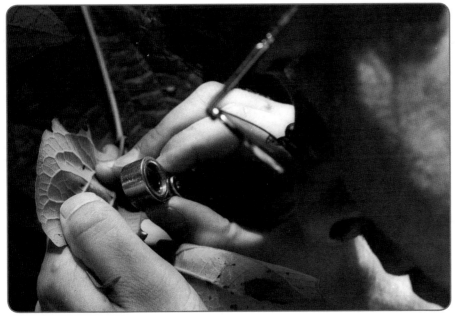

▲ Scientist discovers new species

a Suggest how a scientist would know if they discovered a new species that had not been named by anyone else before.

b Imagine that you have discovered a new species of whitefly. Its first name must be Encarsia but its second name can be anything you like. Suggest a name for this new species: Encarsia ....?

▲ Each of these students has a first and second name

Classifying organisms

Organisms are classified by placing them into different groups. This is how we place animals and plants into the plant or animal kingdom.

It is an animal if:	It is a plant if:
it can move independently	it can only move in response to external conditions
it cannot make its own food	it makes its own food
all its parts are close together to aid movement	it can grow and spread out over a large area as it does not move about
it does not have chloroplasts	it has chloroplasts and is green

▲ *Worms do not have backbones and are invertebrates*

Animals can then be classified into two more groups:

• those with backbones, the **vertebrates**
• those without backbones, the **invertebrates**.

Vertebrates can then be placed into five different groups.

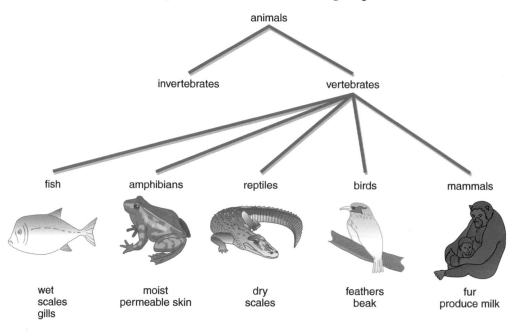

▲ *Snakes do have a backbone and are vertebrates*

▲ *Vertebrate classification tree*

All of these five different groups can be divided into smaller and smaller groups until we get to individual species.

What is a species?

A species is a group of organisms that reproduce with each other. They cannot reproduce with any other organism to produce offspring that are fertile. Members of the same species have more features in common than they do with organisms of a different species.

Even so, members of a species can still show variation. These breeds of dog are all different, but they all belong to the same species.

labrador   poodle   dachshund   Yorkshire terrier   greyhound

Similar habitats

We usually find similar species in similar types of habitat. On grassland in England we find grazing animals such as sheep, cows and horses. In a similar habitat, like the African bush, we also find grazing animals. But this time they are zebras and gazelles.

c In Africa lions feed on grazing animals. In Australia dingoes feed on grazing animals. Suggest why there are no animals that feed on cows and horses in England.

d Suggest what grazing species have evolved in the Australian outback

Different habitats

The apes, for example, consist of many closely related species. Although they are all similar, they are also different from one another because they live in different habitats. The gorillas are big and heavy because they have evolved to live and gather food on the ground. Chimpanzees are smaller and lighter because they have evolved to gather food up in the trees.

Species that are similar to one another, such as the gorilla and the chimpanzee, are usually closely related. They have both evolved from a common ancestor.

▲ Gorilla

▲ Chimpanzee

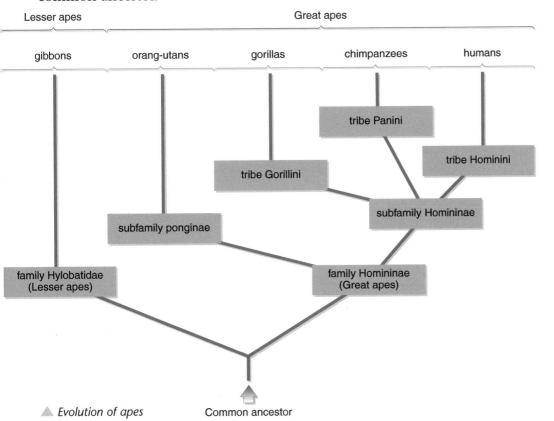

▲ Evolution of apes   Common ancestor

keywords

binomial • invertebrate • vertebrate • species

From chaos into order

Imagine you had a really untidy bedroom and then wanted to find something. You would find it difficult and it would take a long time. It is much easier to find things when they are all tidied away in their own boxes.

In a similar way, classifying or sorting things into groups makes it much easier to identify something. This is why we classify living organisms using the binomial system.

There are many different ways of classifying things. When we do it with living organisms we try to choose those organisms that are closely related to each other and put them into one group.

The organisms that are different are put into another group. Each of these groups is then divided again and again until we have identified each species.

Look at the following animals that you might find in a garden.

| frog | toad | newt | sparrow | blackbird | finch |

Classify the animals by dividing them into two groups. Then take one of those groups and divide it into two more groups. Keep repeating the process until all the objects are classified.

Questions

1 What choice did you make when you divided the animals into the first two groups?

2 How many times did you divide them into groups until they were all classified?

3 Can you think of any other way of classifying the animals? What was your first choice this time?

4 Explain why classifying organisms with a key is much better than comparing organisms with lots of photographs.

Plant magic

In this item you will find out

- how plants make their own food by photosynthesis

- that glucose can be converted into many other substances

- how the rate of photosynthesis can be increased

About 2000 years ago, a Greek philosopher called Aristotle thought plants grew by absorbing all their food from the soil.

If his idea had been correct, then trees and plants would have used up all the soil years ago and simply fallen over.

Aristotle did not know about the different gases in air. Today we know that the air contains **carbon dioxide**.

Plants absorb the carbon dioxide from the air and use it for **photosynthesis**.

This means that the wood we use for tables and chairs is actually made from carbon dioxide and not from the soil in which the tree was growing.

Because it is the leaves that carry out photosynthesis and absorb carbon dioxide from the air, they are sometimes called 'food factories'.

a Where did Aristotle think that plants got their food from?

▲ These trees are made from carbon dioxide

Amazing fact

Almost 40% of photosynthesis on Earth is carried out by plankton in the sea.

Photosynthesis

Plants make food by converting carbon dioxide and water into **glucose** and oxygen. To carry out photosynthesis they need energy from sunlight and a green chemical called chlorophyll. Chlorophyll is found in leaves. This is why nearly all leaves are green.

$$\text{(light energy)}$$
$$\text{carbon dioxide + water} \rightarrow \text{glucose + oxygen}$$
$$\text{(chlorophyll)}$$

▲ *Light and chlorophyll*

 The average tree absorbs about 16 kg of carbon dioxide each year.

A gallon of petrol used in a car produces about 8 kg of carbon dioxide.

If a motorist uses 500 gallons of petrol a year, how many trees are needed to absorb all of the carbon dioxide?

Glucose

The glucose produced by photosynthesis is very soluble and is dissolved in the plant's sap. The plant can then transport the dissolved glucose to any other part of the plant. Very often it is transported to the plant's roots for storage as insoluble starch.

The plants use the glucose for **respiration**. This releases energy for the plants to use. Most people think that only animals respire but in fact plants do as well. Just like animals, they need energy to function. This means they have to respire all of the time.

Other uses for glucose

Plants make much more glucose than they need for respiration. The spare glucose can be changed into **cellulose**, **proteins**, **starch**, **fats** or **oils**. Plants that store food as starch include potatoes, wheat and corn. Bread flour made from wheat is mainly starch.

Some plants, such as oil seed rape, convert and store the glucose as fat or oil.

▲ Flour is mainly starch

▲ Oil seed rape

Proteins are used by the plant for growth and repair. Cellulose is the material that plant cell walls are made from. Humans cannot digest cellulose – when we eat plants, the cellulose forms part of the roughage or fibre that passes straight through our gut.

Making photosynthesis work faster

Plants grow faster in the summer because it is warmer and there is more daylight. They grow faster because they carry out more photosynthesis. It would be useful if we could make photosynthesis work even better so that farmers could grow plants more quickly. Fortunately this is possible.

- We can increase the amount of carbon dioxide that is available for plants to use. Because carbon dioxide is a gas, we can only do this effectively in a greenhouse. Outside, the carbon dioxide would just blow away.

c What effect do you think the increasing levels of carbon dioxide in the atmosphere are having on the rate of photosynthesis?

- We can increase the amount of light reaching the plants. We can increase either the brightness of the light or the number of hours of daylight.
- We can increase the temperature around the plants. Photosynthesis is a chemical reaction. Chemical reactions happen more quickly when the temperature increases.

d State three ways that we can increase the rate of photosynthesis and growth in plants.

The green revolution

Dan Burton works in research and development for a manufacturer of greenhouses. The greenhouses are used by large multinational companies to grow food all around the world. These companies want to grow more food more quickly every year and they want greenhouses that will help them do that.

Dan knows that to increase the rate of photosynthesis requires increasing the amounts of carbon dioxide, light and heat the plants receive. He carries out some experiments. In each experiment one of the variables is changed, while the other two remain constant.

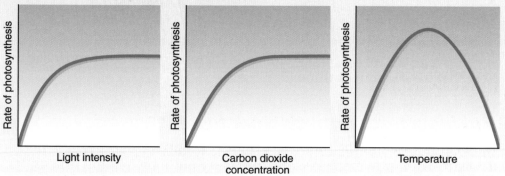

▲ Rate of photosynthesis under different conditions

Graph 1 shows the amount of carbon dioxide against the rate of photosynthesis.

Graph 2 shows the amount of light against the rate of photosynthesis

Graph 3 shows the temperature against the rate of photosynthesis

Questions

1 What happens to the rate of photosynthesis as the amount of carbon dioxide increases?

2 What happens to the rate of photosynthesis as the amount of light increases?

3 Describe how the graph for carbon dioxide and light is different from the graph for temperature.

4 Suggest a reason for this difference.

The fight for survival

In this item you will find out

- how animals and plants compete with each other
- about predators and prey
- about organisms that rely on other organisms

Unlike humans, animals cannot go to the shops to buy food when they are hungry.

Animals must catch and eat food to survive. This means they must **compete** with other animals for the same food supply.

Competition between different animals and plants is a natural way of making sure that the population of one species does not get too big or spread over too large an area. If a population does get too big then a shortage of food or water, or a lack of space or shelter, will bring it back under control.

Amazing fact

A rat can last longer without water than a camel can.

 Name three factors that can affect the size of a population.

Competition ensures that the population level remains fairly constant.

Sometimes this form of control can go wrong. For example, in 1976 there was a plague of ladybirds. The air was full of flying ladybirds, and piles of dead ladybirds were found everywhere. They became a real nuisance.

All the ladybirds were competing for the same food and later the population crashed as they died of starvation. Ladybirds were hard to find for the next few years.

b Explain why ladybirds were hard to find in 1977 and 1978.

Let the best organism win

Each individual organism on the planet is in competition with all the other organisms for survival. Animals compete for food, water, shelter and mates. Plants compete for light, water and minerals. Organisms that fail to compete successfully will die. Only successful organisms will go on to survive and breed. There are rarely enough resources for all the plants and animals. This means that the resources limit how big the population can be and where the population can live.

Habitats can also only support so many different species, particularly if they are competing for the same food, light, water or space. The most successful species survives and the least successful dies.

▲ Hedgehogs are destroying bird life in the outer Hebrides by eating the eggs of ground-nesting birds

▲ The roots of bracken contain poison that kills off other plants

Predators and prey

Predators and **prey** both affect the size of each other's population.

When the population of the predator increases, they eat more prey. This makes the prey population fall. Because there is now less food, the population of the predator falls. There are now fewer predators so the population of the prey increases again.

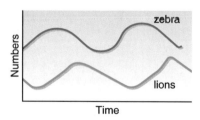

▲ The graph shows the relationship between predators and their prey

c Look at the graph. What do you notice about the number of lions compared with the number of zebras?

d Look at this food chain and state which is the predator and which is the prey.

berries → mouse → kestrel

Living together

Some organisms survive because they rely on organisms from a different species.

Some organisms can only survive by living on or in the body of other organisms. They are called **parasites** and the organisms they live on or in are called **hosts**. Examples of parasites include fleas and tapeworms.

Fleas survive by living on the skin of an animal and sucking its blood. This can weaken the animal and also introduce dangerous diseases into the bloodstream.

Tapeworms grow in the gut of an animal and feed off the food the animal eats. In severe cases they can cause a blockage in the animal's gut. They can grow up to several metres long.

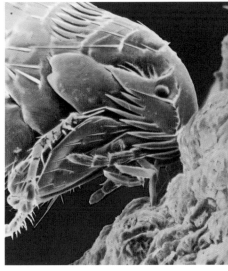

▲ Fleas feed on blood

e **Describe how tapeworms can harm their hosts.**

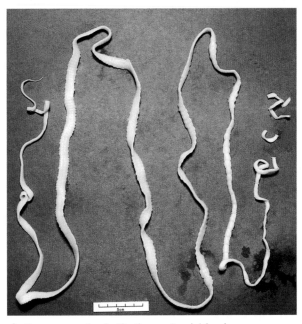

▲ Tapeworms feed off other animals' food

keywords

compete • host
• mutualism • parasite
• predator • prey

Parasites always live at the expense of the host organism. They never do any good and usually harm the host. Some organisms live together so that both of them benefit from each other. This is called **mutualism**.

One example is the oxpecker bird that eats small parasites from the fur of mammals such as giraffes.

Oxpeckers and giraffes benefit from living together ▶

▲ *Red squirrel*

▲ *Grey squirrel*

Illegal immigrants

Red squirrels are native to the UK. This means they have lived here for a very long time.

In about 1850, grey squirrels were introduced into this country from North America. The grey squirrels are more able to compete than the red squirrels.

Grey squirrels do not actually fight with red squirrels, but are bigger and put on a lot more fat than red squirrels. This gives them a much better chance of surviving when food is scarce.

Grey squirrels are also better adapted to live in many more environments than the red squirrel and will happily live in trees in parks, woods, gardens and hedgerows.

Red squirrels are now only found in a few isolated placed in the United Kingdom. You are very lucky if you have ever seen a red squirrel.

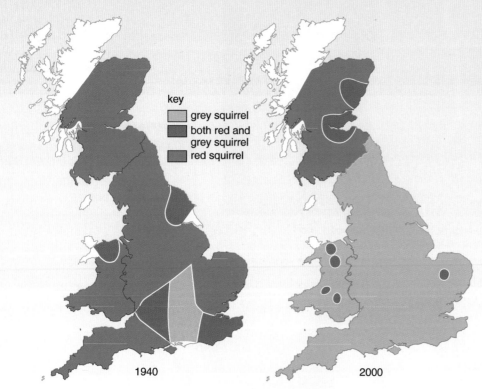

key
- grey squirrel
- both red and grey squirrel
- red squirrel

1940 2000

◀ *Distribution of red and grey squirrels in the UK (1940 and 2000)*

Questions

1 When was the grey squirrel introduced to the UK?

2 Explain why the grey squirrel is more successful than the red squirrel.

3 Suggest where in the UK the grey squirrel was introduced.

4 Suggest how we can ensure that populations of red squirrel continue to survive.

5 Where is the red squirrel most likely to be found today?

Adapt or die

In this item you will find out

- how organisms adapt to their habitats
- how these adaptations help them to survive

Animals **adapt** to suit the habitat in which they live. This helps them compete for limited resources, such as food, water and shelter, for example:

- fish are adapted to swim in water
- birds are adapted to fly in air
- worms are adapted to burrow in the ground.

▲ Cacti are adapted for dry conditions

a Suggest how birds are adapted to fly in the air.

Plants are also adapted to their habitats. For example, cacti and rubber plants are adapted to live in hot, dry conditions such as in a desert.

Amazing fact

The swift is adapted to sleep while flying. This means it can nap between catching insects.

The polar bear

layer of fat for insulation to keep heat in

thick white fur for camouflage and insulation

small ear surface area compared with body, this reduces heat loss

sharp claws and teeth to eat prey

strong legs for running to catch prey

large feet to spread load on snow

fur on soles of feet for insulation and grip

b Suggest why the polar bear needs to be camouflaged.

The camel

fat stored in hump rather than insulating the body

bushy eyelashes and nostrils that can close to stop sand entering

body temperature can increase so it does not sweat and lose water

large feet to spread load on sand

c Suggest why the camel needs to store fat in its hump.

d What feature does the polar bear and camel have in common even though one lives in a hot desert and the other in the cold arctic?

Predator

camouflage so not seen by prey

eyes to the front to judge distance and size

sharp teeth and claws

built for speed

e Explain why the fox has such good ears.

f Explain why the fox has such a good sense of smell.

Prey

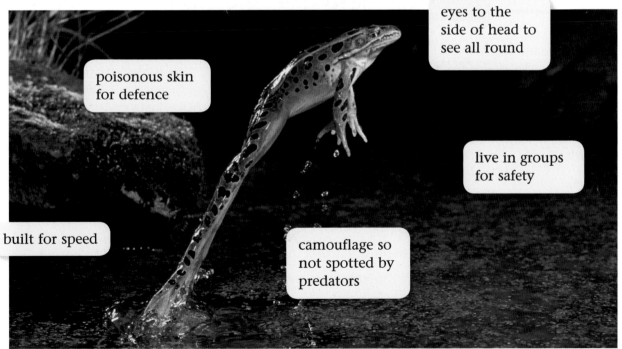

eyes to the side of head to see all round

poisonous skin for defence

live in groups for safety

built for speed

camouflage so not spotted by predators

g What features of a frog help it escape from predators?

Strange adaptions

NEW DEEP-SEA CREATURES FOUND IN ATLANTIC

Some animals and plants have become so adapted to their environments that they are able to live in some very strange places and do some very strange things.

A new type of anglerfish ▶

Scientists have discovered a type of anglerfish never before seen. They are called anglerfish because they attract their prey by waggling a glowing 'lure' attached to the end of a 'rod' on their heads. Anglerfish are found below 1,000 metres in the ocean depths. The new, spiny specimen is thought to belong to the genus Lophodolos.

Scientists said that the structure of its head and luring apparatus is different from the other two known species in the genus.

Some other fish that the scientists caught at the same time were blind. They have lost the use of their eyes because they live in total darkness.

Questions

1 Where did scientists discover the new species of anglerfish?

2 How did the scientists know that they had discovered a new species?

3 How are anglerfish adapted to catch their prey?

4 Suggest what the environment must be like where the anglerfish live.

5 Why are the other fish caught at the same time blind?

All change

In this item you will find out

- how fossils are formed

- how fossils can be useful in understanding evolution

- what happens when environments change

Most scientists believe that the clues to our origin lie in the fossil record. Fossils tell us about plants and animals that lived many years ago.

Fossils found in lower layers of rock are older than fossils found near the surface. This helps us to work out the age of any fossils that we find. Because the age of fossils can be determined, they provide a record of how life has evolved on Earth and how plants and animals have changed over time.

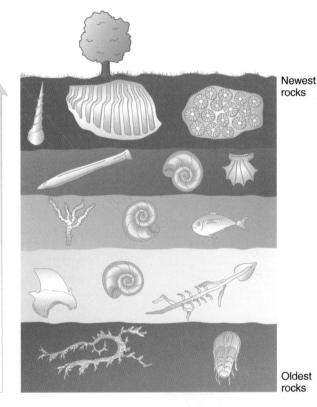

Time

Newest rocks

Oldest rocks

▲ *The younger fossils are near the surface*

How fossils are made

Sometimes when an organism dies it becomes covered with sediment. Over many years the hard parts, such as shells and bones, are gradually replaced by minerals, which then form the fossil.

On rare occasions, the whole organism may be preserved. Examples include insects embedded in amber, dinosaurs that fell into peat bogs or tar pits, or mammoths that became frozen in ice. Even bits of tree trucks can be fossilised and are often found in coal. Animals and plants can also leave casts or impressions which can become fossilised.

We don't have a fossil record of all the plants and animals that ever lived. This is because most organisms do not form fossils when they die. Soft tissue usually decays and does not fossilise. Also, many fossils have not yet been discovered.

Amazing fact

The largest fossil bone ever found was an 8ft long shoulder blade from an ultrasauros.

▲ *Insect in amber*

▲ *Plant fossil in coal*

Why evolve?

Evolution only occurs when the environment changes. Plants and animals that are better adapted to their environments are more likely to survive.

Because all the members of a species are slightly different, some organisms are better adapted to their environments than others.

When the environment changes, some organisms become extinct but some can adapt to the new environment, or evolve. This is called evolution by **natural selection**.

These organisms are more likely to survive and breed. The genes that control these adaptations are then passed on to the next generation.

Natural selection at work

Although natural selection usually takes place over millions of years, it is possible to see it taking place over a much shorter time scale.

During the industrial revolution, heavy pollution covered the trunks of trees and bushes with black soot. Before the pollution the grey speckled peppered moth had excellent camouflage on the bark of the trees. It was very difficult for birds to spot the moths and eat them.

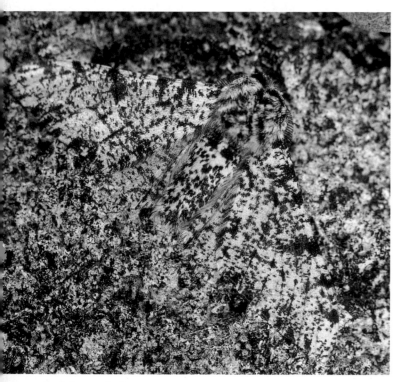

▲ Before pollution

▲ After pollution

When the trees became covered in soot the moths became much more visible and were eaten by the birds.

Some of the moths were slightly darker in colour because of natural variation. These moths had better camouflage against the dark trunks and were not eaten by the birds.

These moths survived and when they bred, they passed on the genes for darker wings to their offspring. Within a few years all the moths were dark and camouflaged.

a What variation in the original peppered moth allowed it to survive in the changing environment?

b What would have happened to the peppered moth if it could not have evolved?

c Suggest what happened to the moths when the trees became clean again.

Bugs and rats

Superbugs are resistant to nearly all of our antibiotic drugs. When antibiotics are used to treat disease, some of the bacteria may be slightly more resistant to the antibiotic.

If these bacteria are allowed to survive, their resistance is passed on as the bacteria multiply. Soon all the surviving bacteria are resistant to the antibiotic.

d Suggest why patients should always complete a course of antibiotics prescribed by their doctor.

Most rats are now resistant to the rat poison called warfarin. This has happened because rats are picky eaters. They taste a small piece of food before returning to it some time later if it has not harmed them.

Those rats that were more resistant to warfarin survived because they only ate a small amount. These rats bred and some of the offspring were even more resistant. After many generations the rats were completely resistant.

So by eating very small amounts of the rat poison that was not enough to kill some of them, they have evolved to become resistant to the poison.

e Explain how rats have become resistant to warfarin.

keywords

evolution • natural selection

▲ After evolution

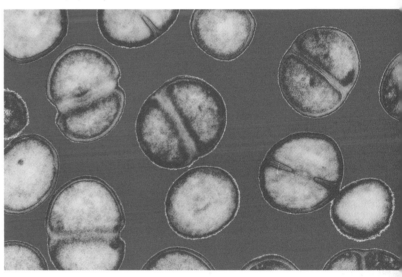
▲ A superbug

▲ Rats are picky eaters in order to survive

Evolution of the horse

The diagram shows how the teeth, legs and body shape of horses have evolved over the past 60 million years.

Look at the data, then answer the following questions.

Questions

1 Describe what has happened to the overall size and shape of the horse over the past 60 million years.

2 How many toes did the horse have 40 million years ago?

3 How many toes does the horse have today?

4 Horses' teeth have ridges for grinding food. Describe what has happened to horses' teeth over the past 60 million years.

5 In which era did Echippus live?

6 How many different varieties of horse have evolved from Merychippus?

Recent

Pleistocene

Piocene

Miocene

Osgocene

Eocone

foreleg

tooth

Equus

3 million years ago

Pichippus

7 million years ago

Merychippus

25 million years ago

Michippus

40 million years ago

Echippus

60 million years ago

▲ Evolution of the horse

Pollution problems

In this item you will find out

- that the human population is increasing

- what the effect this population increase is having on the environment

- how different species can be used to monitor the level of pollution

As the human population increases, more and more of the Earth's limited **resources** of fossil fuels and minerals are being used up. Once used, they are gone forever. Some scientists estimate that within the next ten years, half of all the Earth's crude oil will have been used up.

In parts of the world the human population is growing **exponentially**. This means the population in those places doubles every 53 years. As the population increases, more resources are consumed. This means that we produce more and more **pollution**:

- more household waste
- more sewage
- more sulfur dioxide and carbon dioxide from burning fossil fuels.

▲ Waste disposal tip

Amazing fact

When this sentence was written, the world population was approximately 6,446,131,400. The number is growing by over two people every second.

◄ A modern sewage works can treat over 800 million litres of sewage every day

a By how much will the world's population have grown after just one day?

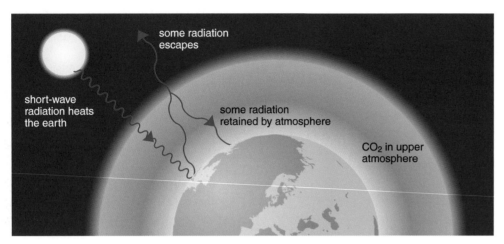

▲ The greenhouse effect

Global warming

When we burn fossil fuels this releases carbon dioxide into the atmosphere. Energy from the Sun hits the Earth's surface and causes warming.

Normally this heat is reflected back into space. Carbon dioxide traps some of this heat energy in the atmosphere just like glass in a greenhouse traps heat. This causes the temperature of the Earth to rise. It is called the greenhouse effect.

b Which gas in the atmosphere is responsible for global warming?

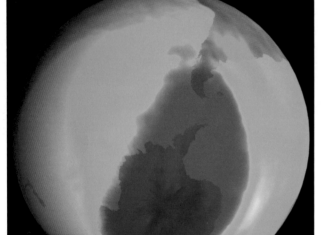

▲ Hole in ozone layer over Antartica

Ozone depletion

The **ozone** layer is a layer of ozone gas that is found in the upper atmosphere. It absorbs most of the ultraviolet light from the Sun. Without the ozone layer, sunlight would contain so much ultraviolet light that it would not be possible for life to exist on the surface of the Earth.

Many of the pollutants that we have produced over the years, such as CFCs from aerosols, have damaged the ozone layer. In recent years, a hole in the ozone layer has appeared over the Antarctic. Fortunately, modern aerosols do not contain CFCs.

c Explain how the ozone layer is being damaged.

d Explain why the ozone layer is so important to us.

This damage was caused by acid rain ▷

Acid rain

Most fossil fuels contain small amounts of sulfur. When the fuel is burnt the sulfur is released into the atmosphere as sulfur dioxide.

Sulfur dioxide dissolves in rain to form **acid rain**. The acid rain kills trees. It also kills fish as it turn rivers and lakes into dilute acid and reacts with limestone buildings and dissolves them away.

Measuring pollution

The numbers and types of organisms that can survive in a particular habitat are affected by pollution. This means that pollution can be measured using biological **indicator species**.

Some fresh water organisms can only survive in clean water with lots of oxygen. If they are present in a river or stream we know the water is clean and pollution-free. These organisms include mayfly and stonefly larvae.

Other organisms can survive in polluted water with very little oxygen. If they are present in a river we know that it is badly polluted.

Organisms found in polluted water include:

• blood worms
• water lice
• rat tailed maggots
• sludge worms.

▲ Blood worm

▲ Water louse

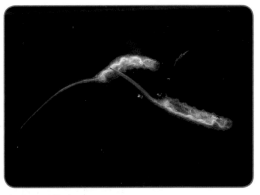

▲ Rat tailed maggot

▲ Sludge worm

Air pollution can be measured using lichens. As air pollution increases, the number of different species of lichen decreases.

◀ This lichen can withstand moderate levels of pollution. See if you can find it on buildings around your school

 Explain how lichens can tell us how much air pollution there is in a particular area.

Testing for pollution

Scientists from the Environment Agency measure the quality of water in rivers and streams.

One way to do this is to catch and examine the types of organism that are living in the stream.

Scientists take a sample of water from the stream and identify and count all of the living creatures that they find.

The numbers of different kinds of organism allow the scientists to work out how polluted the water is.

These results are from a stream that was examined each year for 3 years.

	Number of organisms caught in sample					
	Blood worm	Rat tailed maggot	Sludge worm	Water louse	Mayfly larva	Stonefly larva
Year 1	45	4	8	10	0	0
Year 2	3	1	7	11	1	0
Year 3	0	0	0	7	4	2

Use the information from the table and the previous page to answer the following questions.

Questions

1 Is the stream getting more or less polluted after the three years?

2 Which organisms tell the scientists that the stream is polluted?

3 Which organsisms tell the scientists that the stream is getting cleaner?

4 Which organism is found in both clean and polluted water?

5 Which result from the table may be regarded as an anomalous result?

Extinction is forever

In this topic you will find out

- why animals and plants become extinct
- how endangered species can be protected
- about sustainable development

These are just a few of the species that have become **extinct** over the past few hundred years. They will never be seen on this planet again.

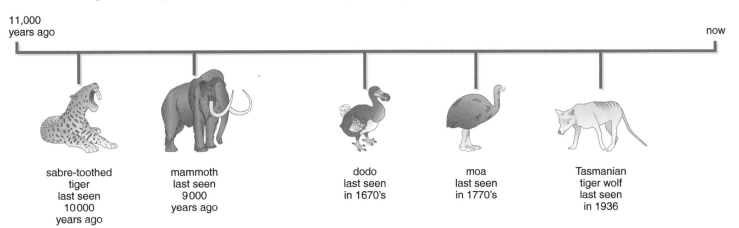

11,000 years ago — now

sabre-toothed tiger last seen 10000 years ago

mammoth last seen 9000 years ago

dodo last seen in 1670's

moa last seen in 1770's

Tasmanian tiger wolf last seen in 1936

Many animals and plants alive today, such as the panda and gorilla, are **endangered** species. This means they are likely to become extinct in your lifetime.

Some organisms become extinct naturally. This can happen when their environment changes due to natural causes such as **climate change**, or it can happen if they have to compete with other more successful species.

 a **Describe two ways that animal extinction occurs naturally.**

Unfortunately, most extinction occurring now is due to the effect humans are having on the environment.

▲ *Giant pandas are an endangered species*

▲ Desmoulins whorl snail

Destroying, hunting and polluting

In 1995 a new bypass was planned around the town of Newbury. The habitat of an endangered rare snail needed to be destroyed to build the road.

The Newbury bypass was built and opened in 1998. Fortunately some of the snails were saved and moved to another site.

The Northern Right Whale is on the verge of extinction. During the 1800s, whaling ships reduced the population from thousands to just a few hundred. They were called Right Whales because they were the right whales for the whaling ships to kill. Whales are hunted commercially for their meat and blubber which is used for food, oil and to make cosmetics.

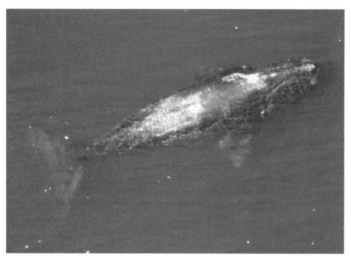

▲ Northern Right Whale

▲ Leatherback turtle

Leatherback turtles have existed for over 100 million years. They are likely to be extinct within the next 10 years. One of the greatest threats to the turtles is eating ocean pollutants such as plastic bottles that are thrown into the sea.

▲ Site of Special Scientific Interest

How can humans help?

It is important that we conserve endangered species and their habitats. They help to provide the rich and varied landscape of our country. Once they are gone it is very difficult to get them back. Some species in Britain, such as the red kite, red squirrel and the osprey, are endangered species and in need of protection.

We can help to protect them in the following ways:

1 **Protecting habitats**
   Habitats that contain rare or endangered species can be protected. They can be labelled as Sites of Special Scientific Interest or SSSIs. This should prevent anyone developing the land and destroying the habitat.

2 Legal protection

Some species are given legal protection so that they cannot be hunted or killed. This law applies to many wild bird species.

3 Education programmes

People can be educated about how fragile and important our environment is. This way they learn how to appreciate and respect it.

4 Captive breeding programmes

Some zoos have breeding programmes to breed rare species and release them back into the wild. Examples include birds of prey such as the red kite and the osprey.

5 Creating artificial ecosystems

Artificial ecosystems can be created for endangered species to live in. Even having your own wildlife pond can make a difference.

b Suggest another artificial ecosystem that could be created to help protect an endangered species.

▲ *The red kite is a protected species*

Sustainable development

One way that the environment can be protected is by **sustainable development**. This means that whatever is removed from the environment is replaced. A resource that is replaced is known as a **sustainable resource**.

Woodland that is cut down for its wood can be sustained by replanting with young trees. The young trees grow and sustain the woodland. In the future, these trees will also be harvested and replaced.

▲ *The osprey is also a protected species*

Many woodlands now have visitor centres that are often used by schools to teach students about conservation, including rare and endangered species.

Fish stocks in the North Sea are being seriously over-fished. The numbers of fish, such as cod, are falling dramatically.

Fishermen have been given quotas to limit the number of fish that they can remove. This should then allow the fish to breed and the fish population to recover.

c Explain how having fishing quotas will allow the fish stocks to recover.

keywords

climate change
- endangered • extinct
- sustainable development
- sustainable resource

▲ *Sustained woodland*

Whale watching

Hamish leads whale-watching tours in the Atlantic. Whales and dolphins are mammals. They are more closely related to humans than other marine animals. Perhaps that is why we have such a fascination with them and whale watching is a valuable commercial tourist industry.

He tells the tour group:

"Different kinds of whales feed on different food so you can see them in different parts of the world.

Whales are still hunted in some parts of the world. They are valued as food and for the oil that they contain. Chemicals from their bodies are also used for manufacturing some cosmetics. This is why some whale species are nearly extinct.

Some whales and dolphins are kept in captivity. They are used for both entertainment and research. Many are bred in captivity and are not used to swimming in the open sea.

Some people think that whales and dolphins should never be kept in captivity.

Other people think that it is OK if they are kept for scientists to find out more about them and how we can help them survive in the wild."

Examiner's tip

This is another area where examiners will expect you to know both sides of the argument. To hunt or to conserve.

Questions

1 Suggest why whale watching is such a popular tourist activity.

2 State two reasons why whales are still hunted by whaling boats.

3 Explain why tourists can see different kinds of whales in different parts of the world.

4 State two reasons why whales are kept in captivity.

5 Although whales and dolphins seem happy to perform for the public, explain whether you think they should be used in this way.

B2a

1 State which apparatus is used with each method of collecting organisms. Choose from this list.

pooter   net   pitfall trap   quadrat

A catching soil insects in the garden
B catching flying insects
C sampling daisies in a field
D catching insects in a sample of leaf litter   [4]

2 Which of the following statements are true about a biological key?

A it can be used to open greenhouse doors
B it can be used to identify organisms
C it works by dividing organisms into groups
D it tells you where different organisms live   [2]

3 State which of the following ecosystems are natural and which are artificial.

A woodland   B fish tank   C desert
D farmer's field   E greenhouse   [5]

4 Describe the difference between an ecosystem and a population.   [2]

5 There are many different types of ecosystem.

a Name an ecosystem that has not yet been fully explored.
b Explain why it has not been fully explored.   [1]

6 Use the key to identify the following organisms.

1 has wings   go to 2
  has no wings   go to 3
2 large antenna   butterfly
  small antenna   housefly
3 has legs   spider
  has no legs   caterpillar   [4]

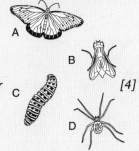

B2b

1 Place the following organisms into their correct group.

rose   fish   daisy   worm   bird   spider

A plant   B vertebrate   C invertebrate   [6]

2 Look at the following animals

State which is a mammal, which is a fish, which is an amphibian and which is a reptile.   [4]

3 Look at the following breeds of dog.

Explain whether the dogs belong to the same or different species.   [1]

4 Describe the characteristics of the following animals.

A fish   B amphibian   C reptile
D bird   E mammal   [5]

5 Explain what the word 'species' means.   [2]

6 Which of the following statements is true?

A Similar species tend to live in similar habitats.
B Similar species tend to live in different habitats.
C Closely related species tend to have similar feature in different habits.
D Closely related species tend to have different features in different habitats.   [2]

B2c

1 Finish the sentences using the following words.

**carbon dioxide   glucose   oxygen
photosynthesis   water**

Plants make food by a process called ____(1).

Plants absorb ____(2) from the air and ____(3) from the soil.

They make the food ____(4) and release a gas called ____(5).   [5]

2 State which of the following plants use for photosynthesis.

A glucose   B light   C sound
D chlorophyll   E haemoglobin   [2]

3 State which of the following statements about plants is true.

A Plants photosynthesise but do not respire.
B Plants respire but do not photosynthesise.
C Plant photosynthesise and respire.
D Plants do neither.   [1]

4 Write down the word equation to describe photosynthesis.   [6]

5 Plants can convert glucose into many other substances. Explain how the plant uses each of the following substances.

A glucose   B cellulose   C protein
D starch and oils   [4]

6 Explain why plants respire all the time but only photosynthesise in the daytime.   [2]

B2d

1 Plants compete with each other for survival.

Which of the following will plants compete with each other for?

A light   B money   C water
D minerals   E space   [4]

2 Look at the food web
a List two predators and two prey. [4]
b Which two organisms can be both predators and prey? [2]

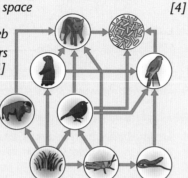

3 Some organisms rely on the presence of a different species. Explain why animals like the buffalo and giraffe rely upon the oxpecker bird. [1]

4 Populations of organisms are not evenly distributed on the planet. List three factors that can affect their distribution. [3]

5 In a predator – prey relationship, describe what will happen when:
a the number of predators increases [1]
b the numbers of prey decreases [1]

6 Finish the sentences using the following words.

mutualism   parasites   prey   predators

Animals that feed on other animals are called ____(1).
The animals that they feed upon are called ____(2).
Some animals live on or inside the animal that they are feeding upon. They are called ____(3).
Sometimes two different species live closely together and depend upon each other. This is called ____(4). [4]

B2e

1 Look at the following organisms.

camel   bird   fish   worm

State which is adapted to each of the following habitats.

A water   B air   C underground   D desert   [4]

2 State which of the following adaptations are often found in predators.

A eyes at front of head   B colourful body
C built for speed   D sharp teeth and claws [3]

3 State which of the following adaptations are often found in prey.

A live in groups   B camouflage
C eyes on side of head   D slow moving   [3]

4 Polar bears are adapted to live in cold arctic conditions. Explain how each of the following adaptations enable the polar bear to survive.

A white fur for camouflage   B layer of fat under the skin
C large feet   D large size   [4]

5 A camel is adapted to live in the desert. Explain how each of the following adaptations enables the camel to survive.

A hump containing fat
B can allow its body temperature to rise
C bushy eyebrows and hairy nostrils
D large feet   [4]

B2f

1 Finish the sentences using the following words.

changed   evolution   fossil   plants

Evidence for ____(1) can be obtained from the ____(2) record.
It shows that animals and ____(3) have ____(4) over time.   [4]

2
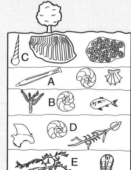

The diagram shows fossils in layers of rock.

a Which fossil is the oldest? [1]
b Which fossil lived for the longest period of time? [1]
c Which fossil is most likely to be alive today? [1]

3 Which of the following statements is true?

A Organisms that are better adapted to their environment are less likely to survive.
B Adaptation does not affect survival.
C Organisms that are less adapted to their environment are more likely to survive.
D Organisms that are better adapted to their environment are more likely to survive. [1]

4 Fossils are the preserved remains of dead organisms.
a Describe three different ways in which fossils can be preserved. [3]
b Give two reasons why the fossil record is incomplete. [2]

5 Which of the following statements are true?

A When the environment changes some organisms may become extinct.

B Evolution only happens when the environment changes.

C Natural selection is when better adapted organisms die.

D Sexual reproduction allows genes to be passed on to the next generation. [3]

6 Put the following statements about peppered moths into their correct order.

A Some moths were slightly darker and had a better chance of survival.

B The industrial revolution produced soot that turned tree trunks black.

C The pale moth could now be seen by predators and was eaten.

D Pale coloured peppered moths were camouflaged on the bark of trees.

E Sexual reproduction produced even darker moths and after several generations all the moths were of the dark variety. [5]

B2g

1 The human population is increasing. Which of the following will reduce as the population gets bigger?

A household waste   B fossil fuels

C carbon dioxide   D minerals   [2]

2 Which of the following will increase as the population gets bigger?

A oil   B sulfur dioxide

C coal   D sewage   [2]

3 Which of the following statements about pollution is usually true.

A Pollution increases the number and different types of organism.

B Pollution decreases the number and different types of organism.

C Pollution decreases the number but increases the different types of organism.

D Pollution increases the number but decreases the different types of organism. [1]

4 Match the following words to the correct description.

carbon dioxide   CFCs   sulfur dioxide

A acid rain   B global warming

C destruction of ozone layer   [3]

5 Look at the graph.

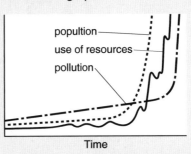

Describe the relationship (pattern) between population growth, use of resources and pollution. [2]

6 The following organisms were found in three different streams.

Stream A – caddis fly and mayfly larva

Stream B – blood worm and rat tailed maggot

Stream C – stonefly larva and water scorpion

Which stream is the most polluted? Explain your answer. [2]

B2h

1 State which of the following animals are extinct.

A red squirrel   B dodo

C red kite   D mammoth   [2]

2 State which of the following animals are endangered.

A panda   B gorilla

C greenfly   D osprey   [3]

3 State which of the following can be a sustainable maintained resource.

A fish farm   B polar ice cap

C desert   D woodland   [2]

4 Match the following reasons for extinction with the correct description.

climate change   habitat destruction   hunting pollution   competition

A fishing for North Sea cod

B two different organisms needing the same source of food

C global warming

D cutting down the rainforests

E the release of sulfur dioxide from fossil fuels   [5]

5 Humans can help endangered species. Which of the following are methods that will be successful?

A legal protection of their habitat

B increase fishing and hunting licences

C captive breeding programmes

D creating artificial ecosystems

E building new roads and towns   [3]

C1 Carbon chemistry

Most of the clothes I am wearing are made of crude oil

Every time my mother fills our car she buys 60 litres of petrol

The newspapers tell us that crude oil is running out. What will we do?

- Every year we hear of new materials and substances that have been developed by chemists. These materials make our lives better.

- Many of these come from crude oil. Polymers such as Gore-Tex®, cosmetics and food dyes all use crude oil as the raw material.

- Exploiting crude oil is important but crude oil is a very flammable substance that has to be handled carefully.

- Crude oil can cause great pollution problems if it escapes into the environment.

What you need to know

- A particle model can be used to explain solids, liquids and gases, including changes of state and diffusion.

- Mixtures are composed of constituents that are not combined.

- How to separate mixtures into constituents using distillation.

Food, glorious food

In this item you will find out

- about different methods of cooking
- what cooking does to the food
- how baking powder makes a cake rise

The next time you go shopping in a supermarket, think about all the different types of foods that are available. The choice is amazing. Look at the foods in all the photographs on this page.

Some of the foods are eaten raw. Some have to be cooked. Some can be eaten raw or cooked.

a Which foods have to be cooked?

b Which foods can be eaten raw?

You cook food for a number of reasons:

- the high temperature kills harmful microbes
- the texture of the food can be improved
- the flavour of the food can be improved
- the food becomes easier to digest.

There are different methods of cooking. These are summarised in the table.

Method	Where	Example
baking and roasting	in a conventional oven	cake, joint of meat
boiling	in a saucepan of boiling water	an egg or potatoes
steaming	in steam above boiling water	fish or vegetables
frying	dipped into heated fat in a frying pan or deep fat fryer	chips, eggs or deep fried fish
grilling	under a grill	fish, steak
by microwave	in a microwave oven	fish, precooked food

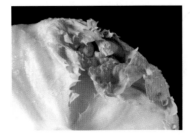

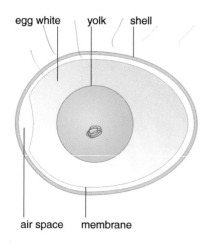

egg white yolk shell

air space membrane

The cooking process

When a chemical change takes place:

- a new substance is formed
- the change cannot be reversed (it is irreversible)
- an energy change takes place.

All of these factors apply when cooking food. Therefore cooking food is a chemical change. So what happens when you cook eggs, meat or potatoes?

An egg is made up of three parts – the shell, the egg white and the yolk. The diagram shows the parts of an egg.

The table summarises some information about an egg.

Part of the egg	Appearance	What it consists of
shell	hard white or brown solid	largely calcium carbonate
white	thick colourless liquid	one eighth protein and seven eighths water
yolk	orange or yellow emulsion	one third fat, one sixth protein and a half water

Egg is a good source of **protein** as it is found in both the yolk and the white.

c Humans need proteins, fats and **carbohydrates**. Which food type is not in an egg?

When you heat an egg the proteins change permanently. The protein molecules change shape.

The photographs show an egg before and after cooking.

▲ *Before cooking*

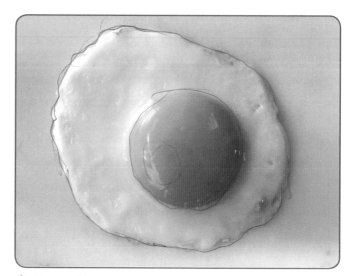

▲ *After cooking*

The change to the egg white is easy to see. It changes from a thick colourless liquid to a white solid as the egg cooks. The changes to the yolk are more difficult to see, but if the egg is overcooked the yolk breaks up into a yellow powder.

Meat is also a good source of protein. The pie diagram shows the food types in a sample of beef.

When meat is heated to 60°C the protein molecules start to change shape. As the meat cooks, its colour usually changes from red to brown and the texture becomes chewier. But these changes depend on how it is cooked.

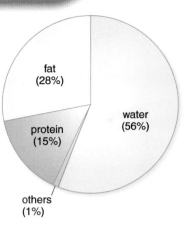

 d Cooking meat can make it shrivel. Suggest why. (Hint: look at the pie diagram)

Potatoes are a good source of carbohydrates. When they are cooked they absorb water. The cells soften so they become easier to eat.

Baking powder

When making cakes, it is important that the mixture rises during baking to give the final product a structure with lots of trapped bubbles. The photograph shows the bubbles in a cake.

e Why is it important to have lots of bubbles in a cake?

When you make cakes you use a raising agent. The simplest raising agent is **baking powder** which contains **sodium hydrogencarbonate**, or bicarbonate of soda. This breaks down or decomposes in the baking process to produce carbon dioxide, water and sodium carbonate. It is the carbon dioxide that makes the cake rise and creates the bubbles.

sodium hydrogencarbonate → sodium carbonate + water + carbon dioxide

Testing for carbon dioxide

You can carry out a test to see if any gas released in a chemical reaction is carbon dioxide.

You bubble the gas through colourless limewater solution (calcium hydroxide). If the gas is carbon dioxide it will turn the limewater cloudy.

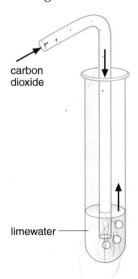

◀ How to test for carbon dioxide

Egg issues

Every day people in Britain eat about 30 million eggs, of which about one sixth are imported.

Most of these imported eggs come from Spain and they are cheaper than British eggs. Spanish eggs are usually bought by caterers.

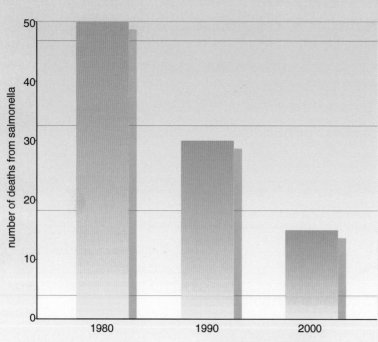

▲ *Deaths from salmonella*

Last year in Britain, there were 80 known cases of salmonella and 15 people died of the illness.

Salmonella is present in raw eggs but it can be destroyed by cooking.

Many of these cases could be traced back to Spanish eggs. A good number of people recover from food poisoning caused by salmonella but certain groups are more vulnerable.

Health authorities are recommending that caterers should not use Spanish eggs in food for older people, the sick and young children.

Also they should not be used in preparing food where the egg is left raw or only lightly cooked.

Questions

1 How many eggs are imported into Britain each day?

2 Suggest why caterers buy most of the imported eggs.

3 Why are infected eggs more of a problem when they are raw or lightly cooked?

4 Which groups of people are more vulnerable?

5 Suggest why these groups of people are more vulnerable?

   The graph shows the number of deaths from salmonella in 1980, 1990 and 2000.

6 What does the graph show about how the number of deaths has changed from 1980–2000?

7 Why should the three points on the graph not be joined with straight lines?

Eating by numbers

In this item you will find out

- about food additives and their advantages and disadvantages
- about food packaging
- about emulsifiers

▲ Froot Loops ®

One brand of breakfast cereal in the USA is called Froot Loops®. The photograph shows a bowl of Froot Loops®.

Like all the food we eat, Froot Loops® cereal is made of chemicals. Each piece of cereal is flavoured and coloured by using other chemicals. These chemicals are called **food additives**.

The different types of food additive and the reasons why they are used are shown in the table.

Food additive	Reason for using it
food colours	to make the food look more attractive
antioxidants	to slow down the food going bad by reacting with oxygen
emulsifiers	to keep the different ingredients thoroughly mixed
flavour enhancers	to improve the flavour of the food

Amazing fact

Each year we each eat 4 kg of food additives in our food.

Hundreds of years ago meat was preserved by adding salt. Salt is also an additive.

 a Suggest what other methods, apart from adding food additives, can be used to slow down the rate at which food goes bad?

Food manufacturers can only add certain approved food additives.

How do we find out what the food we buy contains and which food additives we are eating?

The label on a food packet lists the ingredients in the food. The ingredients are listed in order. The first one in the list is there in the largest amount.

 b What is the main ingredient in the pot rice?

INGREDIENTS AS SERVED (Greatest first)

Rice, Water, Beef, Prepared Soya Protein, Onion, Cornflour, Red & Green Peppers, Tomato, Sugar, Carrot, Peas, Beef Fat with Antioxidant (BHA), Curry Spices, Salt, Yeast Extract, Citric Acid, Flavour Enhancers (Monosodium Glutamate), Colour (Caramel) and Acidity Regulator (Sodium Citrate). Less than 10% meat as served.

▲ The label from a pot rice pot

Approved food additives

All approved food additives have **E-numbers**. The E-number tells us why the additive is used.

E-number	Purpose	Example of use
E100–199	food colours	sweets, soft drinks, jellies
E200–299	preservatives	jams, squashes
E300–399	antioxidants	meat pies, salad cream
E400–499	emulsifiers	margarine and salad cream
E600–699	flavourings and flavour enhancers	sweets, meat products

c **A food label contains the following information:**
Yeast extract, salt, E150c, E220, niacin, thiamin
(i) Why are E150c and E220 added to the food?
(ii) Why do you think salt is not given an E-number?

Food can be coloured by approved food colours, called food dyes. Some of these come from natural sources. For example, caramel made from burnt sugar is used to colour cola drinks. Others come from coal tar. Coal tar is the sticky black liquid produced when coal is heated.

Scientists called food analysts carry out tests on foods. They find out which food dyes have been added to different samples of food, using paper chromatography. It is important that checks like this are made so we can be confident about the food we buy and eat.

The table gives information about four food dyes in jelly babies.

▲ Jelly beans contain food dyes

E number	Name	Colour	Notes
E110	cochineal	red	made from crushed bodies of a type of insect
E120	indigocarmine	blue	made from coal tar
E132	green S	green	made from coal tar (banned in many countries including USA)
E142	sunset yellow FCF	yellow	made from coal tar

d **Which dye in the list is made from a natural material?**

e **The manufacturer is thinking about replacing E132 and using E120 and E142 together to give a green dye. Suggest why.**

Emulsifiers

Oil and water do not mix together. The photograph on the left shows a mixture of vegetable oil and water after it has separated. The oil is floating on top of the water.

▲ Vegetable oil and water

▲ Mayonnaise

It is possible to get the oil and water to mix as an emulsion by adding an **emulsifier** and shaking the mixture. The photograph on the right shows mayonnaise. This is an emulsion of tiny drops of oil spread throughout the water.

An emulsifier molecule has two parts.

These are:

- a water-loving (**hydrophilic**) head
- an oil-loving (**hydrophobic**) tail.

head

tail

▲ An emulsifier molecule

keywords

antioxidant • emulsifier • E-number • food additive • hydrophilic • hydrophobic

Antioxidants

Antioxidants are added to foods to stop them from reacting with oxygen and going bad.

Antioxidants are added to fatty foods such as cakes and biscuits. These spoil when the fats in them are oxidised. Vitamin E is a natural antioxidant. There are some other chemicals that can be used as antioxidants.

Packaging

Using the right packaging can make food last longer and taste better. Active or intelligent packaging can be used to improve the safety or quality of food.

Biscuits are wrapped using a plastic film. This film keeps out oxygen and water which helps to keep the biscuits crisp.

Apples can be wrapped in a special film with very, very tiny holes in it. This helps to keep the fruit fresher longer. Oxygen can enter through these holes.

Some packaging can even help remove water from inside the pack to stop the food going off. Sometimes food is packed in cans that can heat or cool the food inside. You will find an example on page 106.

INTELLIGENT PACKAGING

$E = MC^2$

CRISPS

Should food additives be used?

In 1964, Dr Benjamin Feingold linked the use of food dyes made from coal tar with hyperactivity in children. Hyperactive children become easily excited and have difficulty paying attention. He came to this conclusion after 1,200 experiments. He recommended that these food dyes should be removed from all foods.

E123 (Amaranth) is a purplish-red food colouring. It has been shown to cause birth defects and cancer in rats. Food in the UK can contain E123. Food in Norway and the USA cannot contain E123. At the moment there is no evidence to prove that E123 is harmful to humans.

Scientists test all the chemicals used in food very thoroughly. They repeat the experiments regularly.

Some people think that all natural substances are safe and all manufactured ones are not. MSG (monosodium glutamate) is a flavour enhancer found naturally in seaweed. But if people eat too much MSG it can cause severe headaches.

▲ Brightly coloured jellies contain food dyes

Many people still believe that all food additives are harmful. Some additives have been removed from food after they have been proved to cause dangerous effects.

It can take a long time for any effects to show and then often only in some people. It is not easy to identify which additive is causing the problem.

▲ Monosodium glutamate is found in seaweed

Questions

1 30 years ago more food additives were allowed than are allowed today. Suggest why.

2 Read the paragraph about Dr Benjamin Feingold's work. Why is his work generally accepted by scientists today?

3 Think about the information given about E123. Do you think it should be banned in the UK? Explain why.

4 Why can MSG be a problem for some people?

Heaven scent

In this item you will find out

- about esters and how they can be made
- about solvents
- why cosmetics need to be tested

The cosmetic industry sells millions of pounds worth of cosmetics and beauty products to men and women every day. The photograph shows a girl putting on some make up.

Cosmetics can be made from natural sources or they can be made **synthetically**. Perfumes are often used in cosmetics or they can be used on their own. Natural sources of perfumes include lavender and rose.

What gives perfume its pleasant smell? The smell is due to an organic compound called an **ester**.

Esters can occur naturally, for example the smell of bananas is due to a naturally occurring ester, or they can be made synthetically.

Small amounts of different esters are mixed together to make the perfumes we buy. It takes a skilled person with a very good sense of smell to do this. The mixtures are then tested on consumers to see which are most popular. Many mixtures have to be tried before one is obtained that can be sold commercially.

Perfumes can also be found in many other products such as polishes or foods.

If we are eating a meal, the taste, appearance and smell are all important. Food manufacturers know that adding esters in small amounts can improve the smell of food and make it more attractive.

If you own a shop that sells wrapped bread, you can buy an aerosol that contains a perfume of baking bread, which you can spray in your shop to attract customers.

Amazing fact

One of the most famous perfumes is Chanel No 5. It was first produced in 1921 and sells one bottle every 30 seconds.

 If you go into a department store, assistants often offer to spray a perfume sample onto the back of your hand for you to smell. Why should you leave it a couple of minutes before smelling it?

Making esters

You can make an ester by warming an organic acid with an alcohol. This produces an ester and water. A drop of concentrated sulfuric acid acts as a catalyst.

organic acid + alcohol → ester + water

methanoic acid + ethanol → ethyl methanoate + water

The displayed formula of ethyl methanoate is shown below.

▲ Ethyl methanoate

b Write down the names of the three elements combined in ethyl methanoate.

c What is the molecular formula of ethyl methanoate?

d What is the name of the ester formed from ethanoic acid and methanol?

The table gives the names of some common esters and the acids and alcohols used to make them. It also gives the smell of the ester.

Ester	Alcohol	Acid	Smell
methyl butanoate	methanol	butanoic acid	pineapple
ethyl ethanoate	ethanol	ethanoic acid	pear drops or nail varnish remover
methyl salicylate	methanol	salicylic acid	oil of wintergreen
methyl benzoate	methanol	benzoic acid	marzipan

More about perfumes

We buy perfumes because they have a nice smell.

e A sweetshop sells sweets called pineapple chunks. No pineapple is used to make these. Suggest why these sweets smell and taste like pineapple.

When liquid perfume evaporates, the particles diffuse throughout the room. These particles stimulate scent cells in our noses. This is how we can smell perfume.

▲ Expensive perfumes in a department store

A perfume needs certain properties.

A perfume must:	It must have this property so:
evaporate easily	the perfume particles can easily reach our noses
be non-toxic	it doesn't poison us
not react with water	it doesn't react with our sweat
not dissolve in water	it cannot be easily washed off
not irritate the skin	it can be put directly on the skin

 Explain why it is important that perfumes should not react with water.

Solvents

Have you ever worn nail varnish? When you take it off you use nail-varnish remover. This dissolves the colours in the nail varnish.

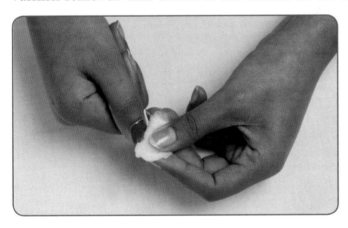

If a substance can be dissolved in a liquid it is **soluble**. If it cannot be dissolved then it is **insoluble**. The substance dissolved in a liquid is known as a **solute** and the liquid which does the dissolving is known as the **solvent**. A **solution** is a mixture of a solute and a solvent that does not separate. Esters can be used as solvents. Ethyl ethanoate is the ester in nail-varnish remover.

 Nail varnish is soluble in nail-varnish remover but insoluble in water. Why is it important that nail varnish does not dissolve in water?

Solvents, such as ethyl ethanoate and hexane, are called non-aqueous solvents. Water is an aqueous solvent.

The table shows which solvents will dissolve wax and salt.

Solvent	Wax	Salt
ethyl ethanoate	dissolves	does not dissolve
hexane	dissolves	does not dissolve
water	does not dissolve	dissolves

h **Copy and complete the sentence by adding either 'aqueous' or 'non-aqueous'.**

Salt dissolves in _____ solvents and wax dissolves in _____ solvents.

---

keywords

ester • insoluble • soluble
• solute • solvent
• solution • synthetic

Testing chemicals

When you buy skin-care or cosmetic products you want to know that they will not cause you any harm. Scientists have to test products and their ingredients very thoroughly to make sure they are safe to use.

Any new skin-care or cosmetic products have to go through a rigorous testing programme before they are released for sale. Some products are tested on animals.

There is a ban on testing ingredients for cosmetics and skin care on animals in the UK. But in the European Union (EU) testing on animals can still take place. Approximately 38,000 animals are used every year in the EU to test new ingredients and products. There is a plan to stop testing these materials on animals in the EU by 2013.

Testing on animals can show scientists how a product or ingredient may affect humans. But sometimes it can be difficult to predict the effect a chemical will have on one species from the test results on another. For example, bleach causes severe irritation to human skin, but only mild irritation to rabbit skin.

Companies in the EU are spending very little money finding alternatives to testing on animals as it is an easy way of getting cosmetics and skin-care products into the shops. There are test-tube methods of testing for possible harmful effects but so far only three of these tests are approved.

Questions

1 Many people do not like the idea of testing on animals. Suggest why a complete ban on testing ingredients and products in the UK and EU would not solve the problem.

2 Why do you think animal tests on some chemicals used in cosmetics and skin care are still allowed?

3 Research on rabbits suggests that a particular chemical will have no effect on humans. Should the scientists accept this research?

4 Why do companies in the EU not spend more money on alternatives to tests on animals?

Cracking good sense

In this item you will find out

- that crude oil, coal and gas are fossil fuels and are non-renewable

- how crude oil can be turned into useful products by fractional distillation and cracking

- about some of the environmental problems caused by the oil industry

Crude oil is a black treacle-like substance usually found inside the Earth. It is often pumped up to the surface. The photograph shows crude oil being poured.

Humans have known about crude oil for thousands of years. It sometimes escapes through cracks in the ground onto the Earth's surface. It is then known as pitch.

Builders of wooden ships such as HMS Victory used pitch. This was to make sure water could not get between the wooden planks.

Crude oil only became valuable when people learned how to turn it into useful products. Crude oil is a **fossil fuel**.

Crude oil was produced over millions of years, through high temperatures and pressures acting on the remains of dead sea creatures.

Coal and natural gas are also fossil fuels. All fossil fuels take millions of years to make. They are called **non-renewable fuels** because we are using them up more quickly than they can be created. Crude oil, coal and natural gas are **finite resources**.

> **Amazing fact**
>
> In Texas 100 years ago, farmers used to burn the pitch on the surface of the Earth because they had no use for it.

 We are being encouraged to use renewable fuels such as wood, straw and sugar rather than non-renewable fuels. Suggest why this is so.

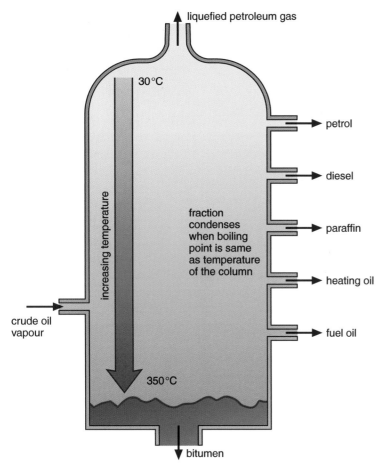

liquefied petroleum gas

30°C

petrol

diesel

increasing temperature

fraction condenses when boiling point is same as temperature of the column

paraffin

heating oil

crude oil vapour

fuel oil

350°C

bitumen

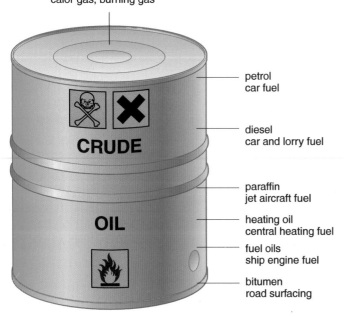

LPG (liquified petroleum gases) calor gas, burning gas

petrol
car fuel

CRUDE

diesel
car and lorry fuel

paraffin
jet aircraft fuel

OIL

heating oil
central heating fuel

fuel oils
ship engine fuel

bitumen
road surfacing

▲ Uses of the different fractions in crude oil

Making crude oil useful

Crude oil is not one pure substance. It is a mixture of many different **hydrocarbons**.

The crude oil can be split into more useful products called **fractions** by **fractional distillation**.

Fractions contain mixtures of hydrocarbons. The process is carried out in an oil refinery. The diagram shows a fractionating column where the separation takes place.

Fractional distillation works because a fraction contains lots of substances with similar boiling points.

Crude oil vapour enters towards the bottom left of the tower. As it passes up the tower, it cools.

A fraction with a high **boiling point** condenses and comes off at the bottom. A fraction with a low boiling point condenses and comes off at the top of the column.

b How is crude oil vapour produced?

c Look at the diagram. Which fraction, petrol or paraffin, condenses at a lower temperature?

The liquefied petroleum gases (**LPG**) do not condense. They contain propane and butane gases.

Supply and demand

The fractions from crude oil are used as fuels. The table shows the supply of each fraction in crude oil and the approximate demand for each fraction by customers.

Fraction	Supply in crude oil (%)	Demand from customers (%)
LPG	2	4
petrol	15	27
diesel	14	21
paraffin	14	9
heating oil	19	14
fuel oil and bitumen	36	25

d Look at the row for petrol. What do you notice about the supply and the demand?

e Look at the row for heating oil. How is this different from the rows for LPG, petrol and diesel?

Cracking

Oil companies can convert large hydrocarbon molecules with high boiling points, such as heating oil, into more useful smaller hydrocarbon molecules with lower boiling points, such as petrol. They can do this by the process of **cracking**. Cracking is done by heating a high boiling point fraction with a catalyst at a high temperature. There must be no air in the apparatus.

 Suggest why there must not be any air in the apparatus.

Cracking also converts large **alkane** molecules into smaller alkane molecules and **alkene** molecules. These alkene molecules are useful because they can be used to make polymers.

The diagram shows apparatus that can be used to crack liquid paraffin in the laboratory. Liquid paraffin contains alkane molecules with about 10 carbon atoms. It is broken down into smaller molecules including ethene.

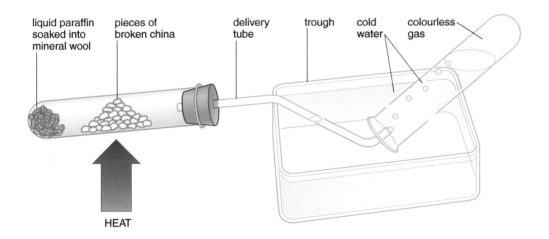

liquid paraffin soaked into mineral wool

pieces of broken china

delivery tube

trough

cold water

colourless gas

HEAT

 During this experiment some oil floats on the surface of the water in the trough. Why is this?

Oil companies often use different cracking processes in the same refinery. Each process uses a different temperature and catalyst. As a result the same crude oil can produce different products.

The table shows the products of two processes, A and B.

keywords

alkane • alkene • boiling point • cracking • crude oil • finite resource • fossil fuel • fraction • fractional distillation • hydrocarbon • LPG • non-renewable fuel

Product	Process A	Process B
methane	13	10
ethene	20	45
propene	12	20
C_6–C_9 hydrocarbons	48	15
other products	7	10

 Suggest why process A is more useful for making petrol, and process B is more useful for making polymers.

Oily problems

12th January 1993

Oil Tanker Causes Environmental Disaster

A week ago the oil tanker MV Braer lost all power. The ship was helpless. It ran aground in hurricane winds on the southern tip of the Shetland Islands. Helicopters rescued the captain and all the crew.

The oil tanker had only a single steel hull. Most oil tankers have a double hull which makes oil leaks less likely. When it ran aground an estimated 85,000 tonnes of crude oil spilled into the sea.

This formed a giant oil slick. On the beaches workers are trying to clear up the oil with detergents and high-pressure hoses.

Seabirds, seals and fish have been badly affected. The oil gets into the birds' feathers so they can no longer fly. When the birds try to clean themselves they swallow the oil and are poisoned. The RSPCA have collected hundreds of dead birds and have managed to clean many others.

Even a week later there are many people trying to prevent this becoming a long-term environmental disaster. There are fears that this spillage will have long-term effects on fish stocks in the area.

MV Braer is a wreck in the bay. You can see the oil slick

Workers are trying to remove deposits of oil from the beach

Questions

1 Suggest how the oil spill could have been prevented.

2 Three things that led to the disaster are mentioned. What are they?

3 The oil slick came ashore in a very beautiful bay. What two effects has this oil slick had on the environment?

4 What long-term effect might this disaster have for the Shetland Islands?

Getting in line

In this item you will find out

- what polymers are
- how to make polymers by polymerisation
- about alkanes and alkenes

The photograph above shows the inside of a modern car.

Many of the things you can see are made from plastics (**polymers**). A plastic is made up of very large molecules called polymer molecules.

 Suggest which materials were used inside cars before plastics were known?

The photograph on the right shows a poly(ethene) filament which forms when two chemicals react together.

Polymer molecules, like poly(ethene), are long chain molecules. In each chain there is a basic unit which repeats itself thousands of times.

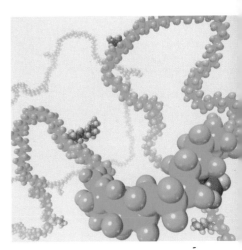

Look at the paper clip in the diagram. It is a single unit. Then look at the chain of paper clips. You can see it is made up of many paper clips joined together.

Just like a chain of paper clips, there is no limit to the number of molecules in a polymer chain.

Polymerisation

The basic unit in a polymer is called a **monomer**. You can make polymers by reacting together lots of monomer molecules. This process is called **polymerisation** and usually needs a high temperature and a catalyst. Poly(ethene) is the polymer made when lots of ethene monomer molecules join together. The diagram summarises the process of polymerisation.

polymerisation continues
until *n* units combined

> **b** Another polymer is poly(propene). What is the monomer for making poly(propene)?
>
> **c** Styrene is a monomer. What is the polymer made from styrene?

Many of these polymer chains jumbled together make up a sample of a polymer.

You can think of the chains in a polymer like the strands in a bowl of spaghetti.

Polymerisation of ethene to produce poly(ethene) can be carried out in different ways to produce different types of poly(ethene).

Polymerisation at a high temperature with oxygen as catalyst produces low-density poly(ethene), LDPE. This is used to make cling-film and polythene bags.

Polymerisation at a low temperature with a different catalyst produces high-density poly(ethene), HDPE. This is used to make milk crates.

Alkanes and alkenes

There are lots of compounds that are made from carbon atoms and hydrogen atoms only. They are called **hydrocarbons**.

d **Here are four formulae. Which one is not a hydrocarbon?**

CH_4 C_6H_{12} CH_2O C_2H_2

Alkanes are one family of hydrocarbons. The simplest alkanes are methane, ethane, propane and butane.

e **What do these four names have in common?**

The diagram shows the displayed formulae of the three simplest alkanes.

methane   ethane   propane

All alkanes contain only single **covalent** bonds between the carbon atoms.

The table shows some information about the first four alkanes.

Alkane	Molecular formula	Melting point in °C	Boiling point in °C
methane	CH_4	−182	−164
ethane	C_2H_6	−183	−89
propane	C_3H_8	−188	−42
butane	C_4H_{10}	−138	0

f **How many atoms are there in a molecule of butane?**

g **What is the molecular formula of the next alkane, pentane, which contains five carbon atoms?**

h **Draw the displayed formula for butane.**

i **How do the boiling points of the alkanes change as the number of carbon atoms increase?**

Alkenes are another family of hydrocarbons. The three simplest alkenes are ethene, propene and butene.

j **What do these three names have in common?**

The diagram shows the displayed formulae of ethene and propene.

All alkenes contain one or more double covalent bonds between their carbon atoms.

ethene   propene

k **Why is there no alkene containing one carbon atom?**

Examiner's tip

Learn the names of the first three alkanes and be able to draw the displayed formulae.

keywords

covalent • hydrocarbon • monomer • polymer • polymerisation

Developing polymers

Polymers are very important materials. Have a look around you and think of how many objects are made from polymers.

Over the years scientists have developed many different types of polymer.

In 1860 there was a shortage of ivory for making billiard balls. John Hyatt produced a substitute called celluloid. This was made from plants using chemicals. It is very flammable. When billiard balls collided quickly the celluloid could produce a mild explosion.

In 1927 poly(vinyl) chloride (PVC) was developed. It is used for raincoats and other waterproof items.

Poly(styrene) was developed in 1930. It is used today for video cassettes and packaging materials.

Eric Fawcett and Reginald Gibson were two British chemists. They discovered poly(ethene) by accident in 1933. It took several years to develop and it played an important role in the Second World War. It is used today for making plastic bags. One type of polythene is LDPE.

In 1935 Wallace Carothers in the USA discovered nylon by accident. It was used to make stockings.

Poly(propene) was developed in 1950. Nine different groups of chemists claimed to have discovered it. After long legal battles over rights, it was decided that two Americans, Paul Hogan and Robert Banks, had discovered it first.

Questions

1 Why did scientists invent celluloid?

2 Which polymers were discovered by accident?

3 Suggest what nylon is used for today.

4 LDPE stands for low density poly(ethene). Suggest what HDPE stands for.

Plastics aplenty

In this item you will find out

- about the uses and properties of some polymers

- how polymers are used in packaging and clothing

- about disposing of polymers

The photographs show some of the everyday uses of polymers. You will be able to think of many others.

a Look at the nylon rope. What properties does nylon need to make it good for making ropes?

b Why do you think PVC is useful for making wellington boots?

The table gives the properties of four polymers.

Polymer	Melting point	Density	Stiff or flexible?
A	high	high	stiff
B	low	low	flexible
C	high	low	flexible
D	low	high	stiff

c A plastic washing-up bowl has to be melted before it can be moulded into shape. It needs to be stiff so it does not change shape. Which polymer should be used?

d A pipe in a car engine has to withstand high temperatures but has to be flexible. Which polymer should be used?

Amazing fact

In 1976 the use of polymers exceeded the use of steel, aluminium and copper combined, for the first time.

Polymers for packaging

Plastic shopping bags are made of polythene. The chemical name for polythene is poly(ethene). The polythene in these bags is very thin but the bags are still strong. Have you seen lots of supermarket bags left about on roads and beaches?

e   **Suggest why people leave these bags around.**

Polystyrene can exist in the form of a foam. Inside the foam are tiny bubbles of air. This foam makes a very good packaging material. It has the advantage of being lightweight.

Polystyrene is also a good insulator. The photograph shows a polystyrene cup. It contains hot coffee but the outside is cool enough to hold.

Plastic clothing

Many of the clothes we wear are made from nylon or polyester.

Nylon has many uses in clothing. It is tough, lightweight, keeps ultraviolet light out and is waterproof. If you are wearing a nylon anorak, what you are wearing underneath stays dry even when it is raining.

But nylon does not let water vapour out. So sweat can condense inside the anorak because it cannot escape. It can be unpleasant to wear a nylon anorak for a long time, especially if you are hot.

In 1958, an inventor called Bob Gore made a material he called Gore-Tex®.

This has all the advantages of nylon but it is also breathable. This means that it allows water vapour to escape from the body.

Clothes made from Gore-Tex® are useful for people who work or exercise outdoors because they stop them becoming wet and sweaty.

◀ *Gore-Tex® is breathable*

Disposal of polymers

Polymers cause problems when they are thrown away in the rubbish or as litter. Most polymers are **non-biodegradable**. This means they do not decay or decompose so they are difficult to dispose of. They can remain in landfill sites for hundreds of years. Landfill sites are filling up and starting new ones wastes valuable land.

◀ *Landfill site*

One solution might be to burn them, but when polymers are burned they can produce toxic fumes. Also, it is a waste of a valuable resource if we dispose of polymers by burning them or throwing them on landfill sites.

These problems could be resolved if more polymers were **recycled**. This is difficult to do because there are so many different types of polymer and sorting them by hand is an expensive process. Mixed polymer waste isn't very-useful. It can only be made into low-value products, such as insulation blocks.

Making biodegradable polymers

The chemical industry is helping the sorting process by stamping marks on items to show what type of polymer they are made from. A bleach bottle has a mark stamped on it to show it is made made from HDPE, high-density poly(ethene).

The future is to produce **biodegradable** addition polymers. Many of these will be made from starch.

It is important that the polymer does not break down too quickly, or it will break down while it is being stored. Starch also swells up when it gets wet. The structure of starch polymers has to be changed to make them water-resistant if these bags are going to be useful for carrying shopping from the supermarket.

▲ *HDPE mark from a bleach bottle*

keywords

biodegradable • non-biodegradable • recycled

Plastic shopping bags in the future

Plastic shopping bags cause lots of problems. They are usually made from polythene, which is not biodegradable. The polythene is made from crude oil and reserves of oil may only last for approximately 80 more years.

Scientists have now developed polymers made from biological materials which are being used to make different plastics.

These bioplastics are a new generation of plastics which scientists claim to be more environmentally friendly than those made from oil.

One problem is that the bioplastics are usually more expensive than the old plastics, but many bioplastics can be processed using ordinary plastic moulding equipment.

Bioplastic bags can be made from starch. This is a renewable carbohydrate polymer that can be purified from various sources by environmentally sound processes. It is found in high amounts in plants like corn (maize), potatoes and wheat.

But starch is water soluble so shopping bags made only of starch will swell and go out of shape when wet. To stop this happening, the starch can be mixed with other materials.

Questions

1 Why do plastic bags cause problems?

2 Why is it necessary to develop plastics that are not made from crude oil?

3 What name is given to plastics made from biological materials?

4 Suggest one advantage and one disadvantage of using these plastics from biological materials.

5 Starch is a renewable polymer. Why is starch renewable?

6 Starch can be used on its own for packaging electrical equipment but not for making plastic bags. Suggest why.

Up in flames

In this item you will find out

- which factors are considered when a fuel is chosen

- about the products of combustion

When fuels burn they give out energy. Imagine that you live in this remote country cottage. It doesn't have a supply of gas or electricity but it could be heated by either coal or oil. Both could be delivered by road.

a Suggest advantages and disadvantages of coal and oil as fuels for your cottage.

This car needs a fuel that can be stored safely in the car. The fuel needs to vaporise (turn to a gas) before it burns in the engine.

b Suggest fuels for the car.

When you are choosing the best fuel, there are seven things you need to think about.

You can remember them if you remember the word **TEACUPS**:

T **toxicity** – how poisonous is the fuel?

E energy value – how much energy does the fuel give out?

A availability – how easy is the fuel to get hold of?

C cost – how cheap is the fuel?

U usability – how easy is the fuel to use?

P pollution – does the fuel produce acid rain or add to the greenhouse effect?

S storage – how easy is the fuel to store?

It has been suggested that hydrogen is the fuel we should use in the future in cars. Why is it a good fuel? It is not toxic. It burns to give out more energy than petrol. It produces no pollution. However, as a gas it is more difficult to store and use than petrol. There is no good source of hydrogen at this time.

Burning fuels

Burning, or **combustion**, of all fuels needs oxygen. The combustion releases heat energy which is very useful. When you burn a hydrocarbon fuel with lots of oxygen (air) you get carbon dioxide and water. This is called **complete combustion**.

The diagram shows an experiment to test what products are produced when you burn methane with lots of oxygen.

The pump sucks air through the apparatus. As the methane burns, a colourless liquid condenses and collects in the first test tube. This liquid boils at 100°C, so it is water.

In the second test tube a gas is bubbled through limewater. The limewater turns cloudy, so the gas is carbon dioxide.

Amazing fact

Coal miners used to detect carbon monoxide in mines by taking canaries with them. The canaries died first if there was gas.

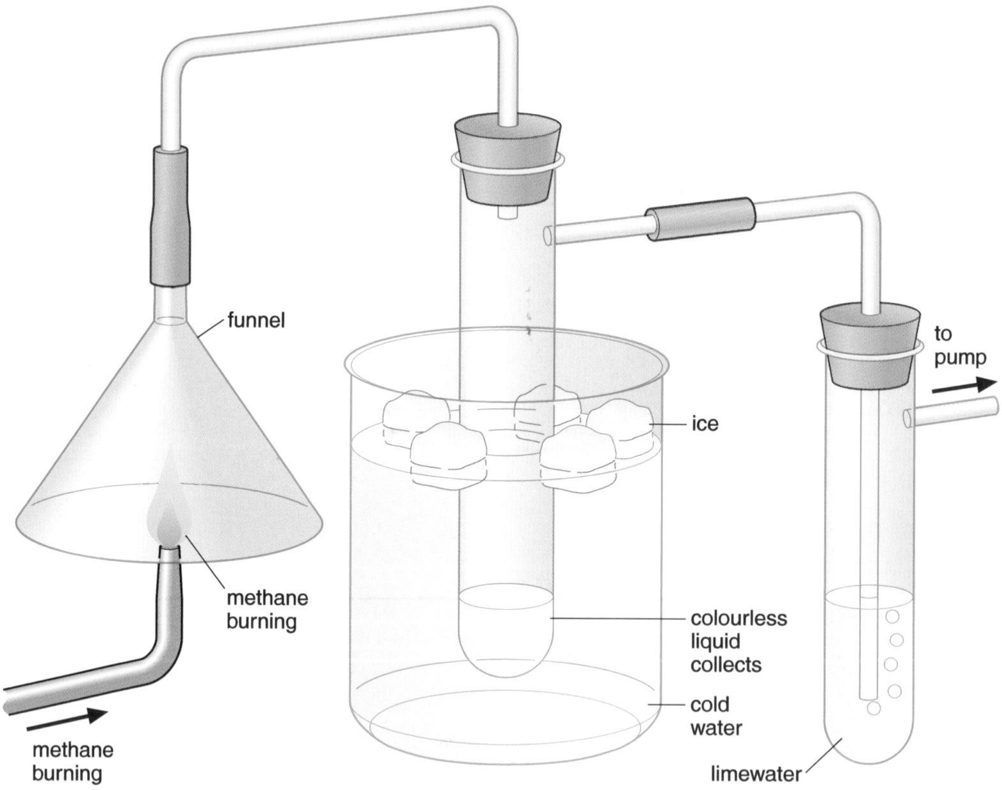

▲ *Burning methane in oxygen*

This can be shown by the following word equation:

methane + oxygen → carbon dioxide + water

Not enough oxygen

If a fuel is burned without enough oxygen then **incomplete combustion** takes place and **carbon monoxide** is produced. This is a poisonous gas.

keywords

carbon monoxide
• combustion • complete combustion • incomplete combustion • pollution
• toxicity

Carbon (soot) and water are also produced. The incomplete combustion of methane can be shown by the following word equation:

methane + oxygen → carbon monoxide + carbon + water

It is better for hydrocarbon fuels to burn with complete combustion than incomplete combustion because:

• less soot is produced
• more heat is produced
• carbon monoxide is not produced.

If you have gas appliances it is important for them to be serviced regularly. If vents get blocked then the fuel could burn with incomplete combustion and carbon monoxide could be produced. In the UK, 50 people die from carbon monoxide poisoning every year.

It is possible to buy cheaply a simple carbon monoxide detector which will detect carbon monoxide in a room for a year. On it there is a cream coloured spot. This goes black if carbon monoxide is detected.

Bunsen flames

The photographs show two Bunsen burner flames. The one on the left is the blue flame and the one on the right is the yellow flame.

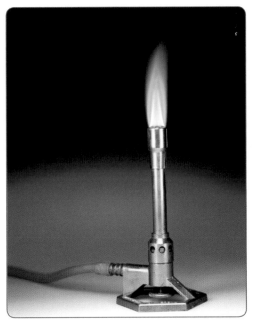
▲ Blue flame

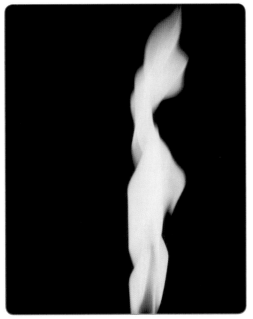
▲ Yellow flame

When we want to heat something to a high temperature we use the blue flame. The blue flame gives out more energy than the yellow flame. This is because it involves complete combustion. The blue flame is also a cleaner flame because it produces less soot than the yellow flame, which produces lots of soot because it involves incomplete combustion.

Choosing the best fuel for a house

Ali and Jo are building a new house. They have to choose the best fuel for their house.

It is hard to compare the relative costs of different fuels as they are frequently changing. Jo and Ali use the Internet to find out the relative costs of fuels. Here are the data.

Fuel	Average system efficiency (%)		Annual running cost (£)
	Room heating	Water heating	
coal	60	25	538
natural gas	70	35	449
LPG	70	35	912
oil	70	35	513

The data can help them decide which fuel they should choose.

Ali and Jo ensure that the builders think about the energy that will be needed to heat the house after it has been built. They could do things during building that could not easily be done later.

Questions

1 Which fuel in the table is least efficient?

2 Are the fuels better at heating rooms or heating water?

3 If Ali and Jo want a cheap fuel, which one should they choose?

4 Which of the fuels in the table do you think Ali and Jo should choose? Their house is not connected to a gas main.

5 Jo suggests they could use electricity to heat the house and produce hot water. She says it is more expensive but produces no pollution. Ali does not agree with this. Suggest why. (*Hint:* think about how most electricity is made.)

6 Jo and Ali intend to live in this house for 12 months before selling it and buying another house. Why may electricity be the best solution?

7 Suggest two things the builders could do to reduce the costs of heating Ali and Jo's house.

8 Suggest why the annual running costs quoted on the Internet for the different fuels are only a guide. The costs will vary from house to house.

Feeling energetic

In this item you will find out

- that chemical reactions can produce heat, light, sound and electricity

- about exothermic and endothermic reactions

- how to compare energy produced by different fuels

Lisa and Jake are sitting round a bonfire.

▲ Burning is a chemical reaction

Even on a cold night the bonfire is giving out a lot of heat. From time to time they are putting on more wood. This is the fuel. The heat energy given out comes from the reaction of the wood with oxygen. They know the more wood they burn, the more heat energy is given out.

Lisa has a torch and a battery radio. She switches on the torch. A chemical reaction inside the torch battery produces electrical energy which is then turned into light. She switches on the radio. Again a chemical reaction in the battery produces electrical energy which is turned into sound.

Lisa and Jake know how important chemical reactions are. They can produce heat energy, light, sound and electricity. But this is only part of the story. Some chemical reactions take in energy. For example, energy is needed for photosynthesis to take place.

I THINK THAT WAS DEFINITELY EXOTHERMIC!

Giving out and taking in

An **exothermic** reaction is a chemical reaction that gives out energy. It releases energy. Burning wood is an exothermic reaction where energy is transferred to the surroundings.

a Can you think of other exothermic reactions?

An **endothermic** reaction is a chemical reaction that takes in energy from the surroundings. It absorbs energy.

You can tell whether an exothermic or endothermic reaction has occurred by looking at the change in temperature that takes place during the reaction.

Experiment 1: A piece of magnesium ribbon is dropped into dilute hydrochloric acid.

The temperature of the solution before adding the magnesium is 20°C. The magnesium reacts and bubbles of hydrogen gas escape. The temperature rises to 26°C.

The increase in temperature shows that it is an exothermic reaction. The increase occurs because heat energy produced by the reaction heats up the solution.

Experiment 2: Some dry sodium hydrogencarbonate and citric acid are put in a test tube. Then some water is added. The temperature of the water is 20°C. There is a reaction and bubbles of carbon dioxide escape. The temperature of the solution is now 16°C.

The decrease in temperature shows that it is an endothermic reaction. The decrease occurs because the reaction takes energy from its surroundings (the solution).

Energy comparisons

When a fuel burns, energy is released. You can measure the amount of energy given out by a fuel by carrying out a **calorimetric** experiment. Look at the apparatus in the diagram opposite (page 105).

The spirit burner contains the fuel. The water is in a copper calorimeter. Before the experiment starts, the mass of the spirit lamp is weighed and the temperature of the water is measured. The spirit lamp is then lit. As the fuel burns, heat energy is transferred to the water and the temperature of the water rises. At the end of the experiment, the flame is put out and the spirit lamp is reweighed. The temperature of the water is also measured.

b James carries out a calorimetric experiment with ethanol. His results are:
temperature of the water at the start = 20°C
temperature of the water at the end = 32°C
What is the temperature change?

c Is this an exothermic or endothermic reaction?

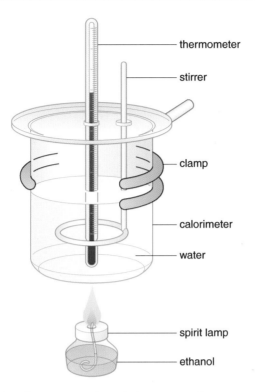

thermometer

stirrer

clamp

calorimeter

water

spirit lamp

ethanol

▲ *Apparatus for calorimetry experiment*

Using this information you can calculate the amount of energy transferred to the surroundings (the water) in units of joules (J) or kilojoules (kJ) where 1 kJ = 1,000 J.

James wants to compare the energy given out by two different fuels. He is going to compare the temperature rise in each case. He is using exactly the same apparatus.

d **Suggest two things he should do to make sure the comparison is a fair test.**

The table gives the results when different fuels burned. In each case 1g of fuel is burned.

Fuel	State of fuel	Energy produced (kJ/g)
hydrogen	gas	243
methane	gas	56
ethanol	liquid	30
petrol	liquid	48

e **Which liquid fuel in the table gives out most energy per g?**

f **Hydrogen could be a good fuel for cars in the future because it causes little or no pollution. What other advantage is there in burning hydrogen?**

The self-heating coffee can

Recently, a new coffee product has been introduced. It is ideal for people who go out for the day into the countryside but like a cup of hot coffee.

All you have to do to get hot coffee is press the plastic button on the bottom of the can.

This starts a chemical reaction inside the can. The energy produced by the reaction heats the coffee in the can.

The coffee can is shown in the picture.

The chemical reaction is between calcium oxide and water. They are stored separately at the bottom of the can.

When the plastic button is pressed, water escapes and comes into contact with the calcium oxide. The word equation shows what happens:

calcium oxide + water → calcium hydroxide

▲ Before the self-heating can!

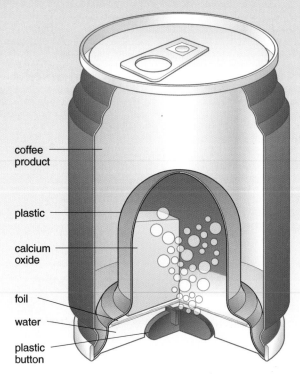

coffee product

plastic

calcium oxide

foil

water

plastic button

▲ How the can works

Questions

1 Suggest what you might do to get a hot cup of coffee if this kind of product was not available.

2 Why is it important that the calcium oxide and water are stored separately?

3 Why is it important that the coffee does not come into contact with the chemicals?

4 What is the product of the reaction?

5 Is the chemical reaction between the water and the calcium oxide exothermic or endothermic? Explain your choice.

6 Are the substances in the equation elements or compounds? Look at each substance in turn.

7 Calcium hydroxide has a formula $Ca(OH)_2$. How many atoms of each element are there in the formula?

C1a

1 Which of the following describe the cooking of an egg? Choose two letters.

A chemical change
B non-reversible change
C physical change
D reversible change [2]

2 Which gas turns limewater cloudy?

A carbon dioxide   B hydrogen
C nitrogen   D oxygen [1]

3 Finish the sentences by using words from the list.

**carbon dioxide   cooled   gas   heated
hydrogen   liquid   rise   sink**

Baking powder gives off a ____(1) called ____(2) when it is heated.

Baking powder makes cakes ____(3). This takes place when the cake mixture is ____(4). [4]

4 Describe the chemical test for carbon dioxide. [2]

5 **a** Finish the word equation for the decomposition of sodium hydrogencarbonate.
    sodium hydrogencarbonate → ____(1) + ____(2) + ____(3) [3]

b Sodium hydrogencarbonate is heated in a test tube. Does its mass increase, decrease or stay the same? Explain why. [2]

6 The formula of sodium hydrogencarbonate is $NaHCO_3$
There are four elements combined.
How many atoms of each element are in the formula? [2]

C1b

1 A fruit drink contains E220 and E163.
Look back at the table on page 80.

What is the job of **a** E220 and **b** E163? [2]

2 Finish the sentences by using words from the list.

**an antioxidant   an emulsion   an emulsifier
a flavour enhancer   a solution**

A mixture of oil and water is called ____(1).
A powdered water-soluble food dye dissolves in water to form ____(2).
MSG is added to food as ____(3).
A substance added to food to stop it from going bad is called ____(4). [4]

3 What are the special features of an emulsifier molecule? [2]

4 Food contains antioxidants.

a What does an antioxidant do in a food? [1]
b Name two foods that contain antioxidants. [2]

C1c

1 Match the words on the left with the correct definitions on the right.

1 solvent   a a substance that dissolves
2 solute   b a liquid that dissolves other substances
3 solution   c a substance that is dissolved
4 soluble   d a mixture formed when substance dissolves in a liquid [4]

2 **a** Suggest why the level of perfume in an open bottle falls quickly if left in a warm room? [1]
b Why does your skin feel cool after putting perfume onto it? [1]

3 Ethanoic acid, CH_3CO_2H and propanol, $CH_3CH_2CH_2OH$ react together to form an ester.

a What conditions are needed to form an ester? [2]
b What is the name of the ester? [1]

C1d

1 A hydrocarbon is:

A a compound containing carbon and hydrogen
B a compound of only carbon and hydrogen
C a mixture of carbon and hydrogen
D an element [1]

2 Identify the four labels on the diagram of the fractional distillation column.

Use words from this list.

**bitumen   crude oil   vapour
heating oil   LPG   petrol**

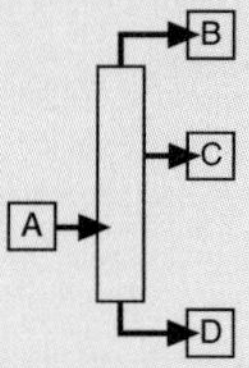

[4]

3 Match the words on the left with the correct definitions on the right.

1 cracking   a turning crude oil liquid into vapour
2 fractional distillation   b producing alkenes from long chain alkanes
3 boiling   c separating different useful products from crude oil

4 Arrange the four fractions in order of increasing boiling point.

fuel oil   heating oil   paraffin   petrol [3]

5 The table contains the boiling temperature range for four fractions from the fractional distillation column.

Fraction	Boiling temperature range in °C
A	70–120
B	120–170
C	170–220
D	220–270

Which fraction comes out highest from the column? [1]

C1e

1 The drawing shows some 'poppet' beads. They join together to form a necklace chain.

a What name do chemists give to a chain of hydrocarbon molecules similar to this? [1]
b Copy the diagram. Draw a ring around a monomer in this chain. [1]
c If each link of the chain represents a styrene molecule, what is the name of the product? [1]

2 The following is a list of hydrocarbons:

ethene   ethane   propane   propene poly(ethene)

a Write down the names of two alkanes. [2]
b Write down the names of two alkenes. [2]

3 Vinegar has a molecular formula $C_2H_4O_2$.

Is this a hydrocarbon? Explain your answer. [1]

4 The linking of small ethene molecules together to form long chains is called _____. [1]

5 The displayed formula for propene is shown below.

Copy the displayed formula. Draw a ring around the feature of the molecule that is typical of all alkenes. [1]

6 The equation for the reaction of propene and hydrogen is:

$$C_3H_6 + H_2 \rightarrow C_3H_8$$

a Draw the displayed formula and name the product. [2]
b What feature in the propene molecule is lost when hydrogen is added? [1]
c What type of reaction is this? [1]

C1f

1 Answer the following questions by using words from the list.

nylon   polyester   poly(ethene)   polystyrene

a Write down the names of two polymers commonly used to make clothing. [2]
b Write down the name of a polymer used to make plastic supermarket bags. [1]

2 a Write down one advantage of waterproof clothing. [1]
b Write down one advantage of breathable clothing. [1]

3 Which two of the following statements explain why more polymers are thrown away and not recycled?

A They are made from crude oil.
B People do not care about recycling.
C Different polymers are difficult to separate.
D They are non-biodegradable. [1]

4 Here is some information about four polymers A–D.

Polymer	Stretches	Easily coloured	Cost
A	easily	yes	high
B	hard to stretch	no	high
C	easily	yes	low
D	hard to stretch	no	low

Which polymer would be best for making bags to wrap vegetables in a supermarket? [1]

5 The table gives information about three materials.

Material	Absorbs water	Sweat absorbed	Breathable
cotton	yes	absorbed	no
nylon	no	not absorbed	no
Gore-Tex®	no	escapes through material	yes

Which material is best for making an anorak for a mountain climber?

Explain your choice. [2]

6 Most polymers are non-biodegradable. Suggest two disadvantages of dumping polymer waste in land-fill sites. [2]

C1g

1 The diagram shows a candle flame.

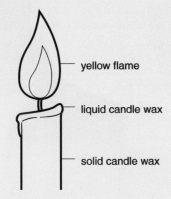

- yellow flame

- liquid candle wax

- solid candle wax

a What is the fuel in the candle? [1]
b Is the combustion complete or incomplete? Explain your answer. [1]
c What happens if a candle is used to heat a glass beaker? [1]

2 Mrs Smith tells her class to leave Bunsen burners on a yellow flame when they are not using them. What is the safety reason for this? [1]

3 Finish the sentences by using words from this list:

fuel   harmless   limited   plentiful   poisonous

Carbon monoxide is a ____(1) gas. It is formed when a ____(2) is burned in a ____(3) supply of air. [3]

4 Raj uses the apparatus on page 100 to show that carbon dioxide and water are formed when a candle burns. He connects the cooled tube and the limewater the wrong way round.

Are his conclusions still valid? Explain. [2]

5 Landlords renting flats to students must have a certificate each year to show that all gas appliances have been serviced. Why is this? [2]

6 A power station in East Anglia is using straw as a fuel to produce electricity. The straw is left over when wheat is grown.

a Suggest an advantage and a disadvantage of using straw as a fuel. [2]
b Why do environmentalists think that burning straw is better than burning natural gas? Ignore the fact that natural gas is a fossil fuel. [2]

7 Write word equations for complete and incomplete combustion of propane. [4]

C1h

1 In any endothermic reaction:

A the temperature falls
B the temperature rises
C the temperature remains unchanged
D there is a change in colour. [1]

2 Look at the diagrams which summarise a reaction.

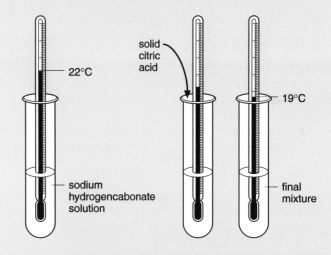

solid citric acid

22°C

19°C

sodium hydrogencabonate solution

final mixture

What can you conclude about the reaction? [1]

3 The word equation for the process of photosynthesis is shown below.

carbon dioxide + water + energy →
   glucose + oxygen

a Is this reaction endothermic or exothermic? Explain your answer. [2]
b Finish the word equation for respiration where glucose and oxygen produce carbon dioxide and water. Include + energy in your equation. [2]

C2 Rocks and metals

Are there paints that change colour?

- Have you ever watched the construction of a large building or perhaps an estate of houses? There is a lot of planned activity.

- Construction materials have to arrive on site at the right time and be of the correct type. Skilled people have to operate expensive machinery.

- All workers have to deliver the building that is on the plan and overcome any problems that occur.

Can we forecast when an earthquake or eruption occurs? Why do they happen?

Why is my asthma worse at some times and in some places?

What you need to know

- The possible effects of burning fossil fuels on the environment.

- The way rocks are formed and destroyed.

- Air is a mixture of gases.

Colour the world

In this item you will find out

- about different types of paint

- about dyes for fabrics

Arthur is a painter and decorator. He uses paints every day. He knows which paints to use for each job.

Over the years the paints have improved. They are easier to use and dry faster and come in many colours.

Arthur knows that is it not a good idea to use cheap paint. It will not look as nice and will not last as long.

Paint can be used to decorate surfaces but it can also be used to protect them from damage or rusting.

 Yesterday Arthur was painting iron railings. Painting makes them look better.

What other reason is there for painting them?

Paints are coloured because they contain **pigments**. Paints also contain a solvent and a binding medium. The solvent thins the paint while the **binding medium** hardens as it reacts with oxygen.

Arthur knows that you do not need to stir most paints before you use them. This is because paint is a mixture called a **colloid**, where solid particles are mixed with liquid particles but are not dissolved.

The pigment particles and binding medium particles are dispersed throughout the liquid solvent particles. These solid particles are small enough so that they do not settle to the bottom. So the paint does not need to be stirred.

Amazing fact

The insides of the tombs in the Valley of the Kings, in Egypt, were painted over 3000 years ago. The paint is still in good condition.

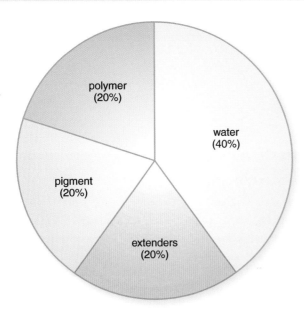

▲ Composition of water-based paint

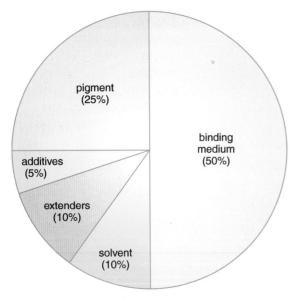

▲ Composition of oil-based paint

Water or oil?

Most paints are applied as a thin surface which dries when the solvent evaporates. Different paints have different solvents.

Emulsion paints are water-based paints. They consist of tiny drops of a liquid polymer (the binding medium) spread out in water (the solvent). The pie diagram shows the typical composition of emulsion paint. The extender is a chalk-like substance which is added to reduce the cost of the paint.

After the paint is applied, the water in the paint evaporates and the polymer particles fuse together to form a continuous film.

Oil-based paints use oil as the solvent and consist of pigments spread throughout the oil. They often contain an extra solvent that dissolves the oil to form a solution. The diagram shows what is in a household gloss paint, which is a type of oil paint.

After the paint is applied the solvent evaporates and the oil hardens to form a paint film.

b Why should a household gloss paint be used in a well-ventilated room?

Special paints

There are now special paints available. Some paints contain pigments that change colour as the temperature changes. These pigments are called thermochromic pigments and the paints are called **thermochromic paints**. The paints could be used to coat a cup. The colour of the cup would change when hot water was poured into it. They could also be used on the outside of a kettle.

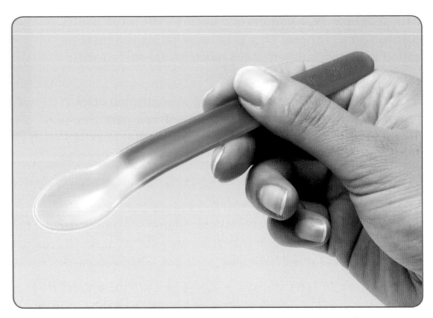

▲ The end of the spoon has changed colour after being placed in hot water

c Suggest why it would be a good idea to have a kettle coated with a thermochromic paint.

There are other paints that glow in the dark. They contain **phosphorescent pigments**.

These pigments absorb and store light energy and then release it over several hours. This means you can see the pigments in the dark.

The photograph shows a watch with phosphorescent paint on the watch face.

d Why would paints containing phosphorescent pigments be useful for seeing signs at night on country roads?

◀ *Phosphorescent paint glows in the dark*

Colouring fabrics

Fabrics can be coloured with dyes. For thousands of years people have used dyes made from plants and other organic materials to colour fabrics. These are called **natural dyes**. For example, if you boil cloth with onion skins they colour the cloth yellow. The dyeing industry took a great step forward when **synthetic dyes** were discovered over 100 years ago. This is because synthetic dyes are available in a wider range of colours than natural dyes.

Traditionally, jeans made from denim were hard wearing trousers for use by cowboys in the USA. They were dyed with the natural blue dye, indigo.

Today many people wear jeans. Some people prefer their jeans to have a worn 'stonewash' appearance. This can be achieved by rotating the fabric in a drum along with some stones. Hence the name stonewash. The stones break some of the fibres and release the indigo dye making the jeans look faded.

Another way of fading jeans is to use the enzyme cellulase as a catalyst. The dyed material is soaked in the enzyme solution. The enzyme catalyses the breakdown of the cellulose fibre releasing some of the indigo dye.

DYEING NATURALLY

Choosing paint

Susan wants to choose a new paint for her bedroom so she goes to Decorama which is a large DIY shop specialising in decorating materials. It sells a wide range of branded and unbranded paints.

The table shows the composition of two brands of gloss paint – one is a famous brand and the other is Decorama's own brand. One is labelled X and the other Y

Components	Percentage composition	
	X	Y
binding medium	54	45
pigment	25	?
solvent	17	20
extenders	0	14
additives (to speed up the drying process)	4	1

Questions

1 Which paint would you expect to dry faster? Explain your choice.

2 Which would be the runnier paint? Explain your answer.

3 What is the percentage of pigment in paint Y?

4 The Decorama paint sells for a cheaper price than the branded paint. Which paint do you think is the Decorama own brand. Explain your reasoning.

5 Suggest one reason why the cheaper Decorama paint may not be the best buy for Susan.

Building basics

In this item you will find out

- about materials used in construction

- about the problems caused when rocks are dug out of the ground

- how limestone can be used to make cement and concrete

Buildings are everywhere but have you ever thought what they are made from? For example, a school is made from lots of different construction materials.

Limestone, **marble** and **granite** may also be used as building materials or as worktops or other fittings.

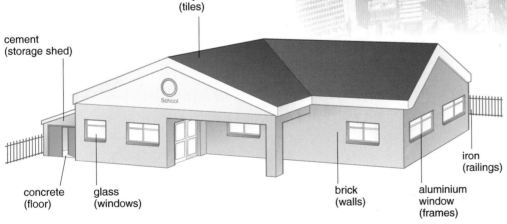

clay (tiles)

cement (storage shed)

concrete (floor)

glass (windows)

brick (walls)

aluminium window (frames)

iron (railings)

▲ What a school is made from

Some construction materials in the diagram are made from rocks taken from the Earth's crust. Iron and aluminium are extracted from rock **ores**. Bricks are made from clay and glass is made from sand.

Today, many houses in the UK are built using bricks. Bricks are all the same size and this makes them easier to use for building than stone.

a What other construction materials come from rocks?

Getting rocks out

Granite, limestone and marble are three rocks used as construction materials. The photographs show samples of each of these rocks.

▲ Granite

▲ Limestone

▲ Marble

▲ A limestone quarry

Granite is much harder than marble. Marble is much harder than limestone.

The photograph on the left shows how limestone is taken from the ground in a limestone **quarry**.

The limestone in the quarry is broken up using explosives and then taken away in large lorries. We get granite and marble from similar quarries.

Getting rocks from a quarry can cause environmental problems. Quarries or mines can take up a lot of valuable land-space. Often quarries are in beautiful landscapes which are destroyed by the quarrying.

Quarrying can also destroy the environment of plants and animals. The landscape then has to be restored when the quarrying is finished which is expensive. There is a lot of dust and noise around a quarry and an increase in traffic on the roads.

More about limestone

Limestone and marble are both forms of calcium carbonate, $CaCO_3$.

b **What are the three elements combined in calcium carbonate?**

Thermal decomposition is a reaction where one substance is chemically changed by heating, into at least two new substances.

Calcium carbonate splits up (thermally decomposes) to form calcium oxide and carbon dioxide when it is heated.

calcium carbonate → calcium oxide + carbon dioxide.

Most limestone is processed to make other construction materials. It can be made into **cement**. To do this, limestone and clay are heated together. The product is then crushed.

When cement is mixed with sand and water, it sets hard. It is used to stick bricks together when building houses and so on.

Cement can also be used to make **concrete**. Cement, sand, gravel (small stones) and water are mixed together and allowed to set. Concrete is a very useful construction material for making railway sleepers, lamp-posts and buildings. It is not very strong however, unless it is reinforced. Reinforcing the concrete with steel rods produces a **composite** material which is stronger than concrete alone.

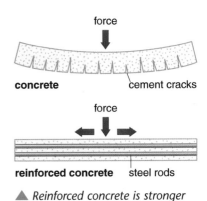

▲ Reinforced concrete is stronger

▲ The wall is being demolished revealing the steel rods used for reinforcement

Choosing a new kitchen

Mr and Mrs Singh are having a new house built. The builder asks them what materials he should use in the kitchen for the floor and the worktops.

They look at different possibilities, but decide they would like stone worktops and a stone floor. But, Mr and Mrs Singh are on a tight budget and they don't want to spend more money on the kitchen than they have to.

The builder gives them some options.

Type of rock	Relative cost	Properties	Does it mark or scratch?	Finishes available
granite	very expensive	very hard	does not scratch or mark	high polish, matt or textured
marble	moderately expensive	not as hard as granite	marks easily but does not scratch	variety of colours, can be polished
limestone	inexpensive	softer than marble or granite	marks easily but does not scratch	cannot be polished

The builder estimates that the area of the worktops is 5 m² and the area of the floor is 24 m².

Questions

1 Which of the three stone materials should they use for the worktops? Explain your choice.

2 Which of the three stone materials should they use for the floors? Explain your choice.

3 Suggest what materials apart from granite, marble and limestone could be used as worktops or flooring? The materials need not be stone.

Restless Earth

In this item you will find out

- about the structure of the Earth
- about tectonic plates
- what causes earthquakes and volcanoes

We are always seeing reports about earthquakes and volcanoes in newspapers and on television. Often an **earthquake** or an eruption of a **volcano** happens unexpectedly.

The photograph above shows an aerial view of the city of Bam in Iran the day after an earthquake on 26 December 2003. It destroyed approximately 60% of the buildings in Bam. Officially, 26,271 people were killed.

The photograph below shows the eruption on a volcano in Bali in Indonesia. Molten lava from inside the Earth spewed out of the volcano. Toxic and unpleasant gases escaped as well.

Sometimes an earthquake underneath the sea can cause a large tidal wave called a tsunami.

To understand these things better we need to know more about the Earth and its structure.

Some people used to think that the Earth was like a solid rubber ball with the same material throughout. We now know that it isn't.

Amazing fact

The Earth has a diameter of about 12,700 km. The deepest that man has been able to drill into the Earth is 15 km.

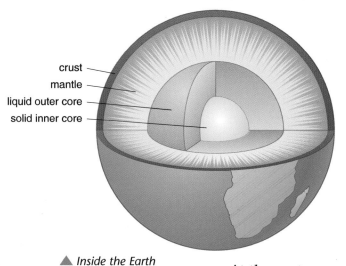

crust
mantle
liquid outer core
solid inner core

▲ *Inside the Earth*

Structure of the Earth

What would you find if you drilled to the centre of the Earth? You would start by burrowing through the rocky **crust**. The thickness of the crust varies from 10 km under the oceans to 65 km under the continents. When you passed through the crust you would notice a rapid rise in temperature.

After the crust you would pass through the mantle – all 3000 km of it. At first the **mantle** is partially fluid but it soon becomes solid rock. This rock is denser than the crust. It would be getting hotter and the pressure would increase from all the rocks above.

At the centre you would see that the **core** is very dense, liquid iron on the outside but solid iron at the centre. The temperature is about 7000 °C but the high pressure keeps the iron solid.

There are problems when studying the structure of the Earth because it is not possible to drill into it at any depth. The only information we have is based on indirect methods. These include the study of different waves that pass through the Earth after an earthquake.

 Would you choose to drill through the crust under the ocean or under the continents? Why?

Tectonic plates and earthquakes

The rigid outer layer of the Earth is called the **lithosphere** and consists of the crust and the upper part of the mantle. It is made up of a number of large sections called **tectonic plates**. The oceanic plates are under the oceans and the continental plates make up the continents. These plates float on the mantle because they are less dense than it. They are moving a few centimetres each year. The diagram shows the plates that cover the Earth.

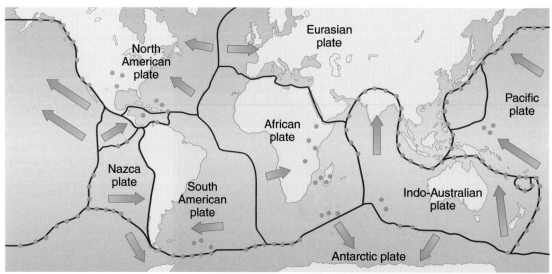

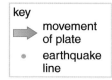

North American plate

Eurasian plate

African plate

Pacific plate

Nazca plate

South American plate

Indo-Australian plate

Antarctic plate

key
→ movement of plate
• earthquake line

▲ *The plates of the Earth*

You will notice the orange dots on the map. These represent places where earthquakes occur. Earthquakes occur where two plates join. The sliding of one plate against the other builds up stresses and strains. When these become too much the result is an earthquake.

b **Look at the map. Do all earthquakes happen where two plates join?**

The movement of tectonic plates can also result in volcanic activity.

Igneous rock and volcanoes

Where there are weaknesses in the Earth's crust, such as cracks or volcanoes, molten rock can make its way through to the surface. The molten rock which is found beneath the surface of the Earth is called **magma**. The magma rises to the surface, usually because it is less dense than the mantle. But if there is enough pressure the magma can rise even if it is denser than the mantle. The molten rock that erupts from a volcano is known as **lava**. Sometimes runny lava erupts from a volcano and sometimes thick lava erupts violently.

Many of the rocks on the Earth's surface have been formed by volcanoes. Molten rock cools and crystallises on the surface to form **igneous rock**. If the crystallising takes place quickly small crystals are formed. The rock can be basalt or its coarse equivalent gabbro. Igneous rocks can also form by crystallising inside the Earth's crust. Here the molten magma cools slowly and large crystals form. This rock can be rhyolite or its coarse equivalent granite. Rhyolite contains a large amount of silica.

Many people choose to live near volcanoes even though there is the risk of eruptions. This is because the soil near a volcano is very fertile so lots of crops can be grown.

Geologists make a special study of volcanoes. They do this because it is the best way of looking inside the Earth without drilling. They are also trying to predict when the next eruption might occur.

▲ Granite

▲ Basalt

keywords

core • crust • earthquake • igneous rock • lithosphere • lava • magma • mantle • tectonic plates • volcano

◄ Cultivating close to a volcano

Volcano watching

Geologists are interested in studying volcanoes. Looking into a volcano enables them to look into the crust of the earth. Volcanoes can erupt in different ways. Some volcanoes erupt explosively, while others erupt slowly over a long period of time. Geologists and other seismologists are constantly taking measurements and observations to try to predict future volcanic eruptions. Their results also give information about the structure of the Earth.

Mount St Helens in the US is monitored regularly to make sure that people can be warned if it is going to erupt again. But the US Geological Survey says that not all active volcanoes that could be a threat are being monitored.

The US Geological Survey suggests that a National Volcano Early Warning System is set up. Scientists would install equipment at volcanoes, which would detect the release of volcanic gases or the swelling of a volcano's surface. The data would be sent by satellite or telephone to scientists for analysis.

Setting up this kind of early warning system for volcanic eruptions is very expensive and requires experienced scientists. Even the best systems cannot be certain exactly when any volcano will erupt.

Questions

1 What name is given to scientists who study the nature of rocks and the structure of the Earth?

2 What name is given to scientists who study the shockwaves that pass through the Earth?

3 Why should volcanoes be monitored?

4 Explain how the National Volcano Early Warning system would work.

5 What other information can studying volcanoes give scientists?

6 There are monitoring stations for earthquakes in the US but not in Indonesia, where there are many more active volcanoes. Suggest why.

Make mine metals

In this item you will find out

- about extraction of copper from its ore
- how copper is purified
- about alloys

◀ *In the picture you will be able to identify many items made of metal*

a How many objects made of metal can you find?

Most metals are extracted from ores. The photograph on the right shows a lump of green malachite. This is a copper ore. The ore is sometimes used in jewellery.

Copper is an expensive metal. This is because there are many uses for copper but very little copper ore on the planet. It is likely to run out in the future.

In order to keep copper available for our children we need to:

- look for new supplies of copper ore
- find alternative materials to copper
- recycle as much copper as possible.

Recycling copper from electrical wires and water pipes, for example, is cheaper than extracting copper from copper ores. Recycling also makes our scarce resources last longer. However, recycling has to be organised so that people know where copper can be collected. People see copper everywhere and may not realise how scarce the resource is.

Amazing fact

Scientists have recently found a polymer called poly(ethyne) that conducts electricity. Perhaps one day this will replace copper wiring.

Extracting and purifying copper

So how do you extract copper from its ore? Copper is a metal which is relatively easy to extract. One method that can be used is to heat a copper ore, such as malachite, with carbon. This is called reduction with carbon.

Copper must be very pure if it is to be used to make electricity wires – at least 99.95% pure. Any impurities will increase the resistance of the wire. Copper can be purified by **electrolysis**. The apparatus is shown in the diagram.

During electrolysis, copper forms on the cathode (negative electrode). In industry this is carried out on a very large scale. The photograph shows a factory where purification of copper is taking place.

▲ Purification of copper on a large scale

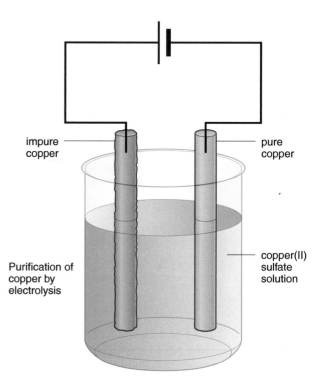

impure copper

pure copper

Purification of copper by electrolysis

copper(II) sulfate solution

Examiner's tip

Remember metals such as copper form on the negative electrode during electrolysis.

The rods of pure and impure copper are called electrodes. The pure copper rod is attached to the negative terminal of the battery. It is called the negative electrode. The impure copper rod is attached to the positive terminal of the battery. It is called the positive electrode. The changes that take place during electrolysis, take place at the electrodes.

Alloys

We don't use very many metals in their pure form. Metals are more useful to us when they are made into alloys. An **alloy** is a mixture of two elements where at least one element is a metal.

◀ *A bronze statue*

The table gives information about some common alloys.

Alloy	Main metal	Use
amalgam	mercury	tooth fillings
brass	copper and zinc	hinges, screws, ornaments
bronze	copper and tin	statues
solder	tin and lead	joining metals
steel	iron	car bodies, ships

The properties of an alloy are different to the properties of the metals it is made from. This often makes them more useful than the pure metals. For example, the bronze in a statue is much harder than the copper and tin used to make it. This makes it more suitable for use as a statue than the pure metals.

The table gives some properties of solder and the metals tin and lead that are mixed together to make it.

Metal	solder	tin	lead
Melting point (°C)	110-130	232	328
Density (g/cm^3)	10.3	9.3	11.3
Relative hardness	quite hard	soft	soft

Solder can be used to join two pieces of copper wire together. The two pieces of wire are placed close together and a piece of solder is held between them. A hot soldering iron is then used to melt the solder. On cooling, the two wires are joined by the solder.

b   **Why is solder better than tin or lead for this purpose?**

keywords

alloy • electrolysis

Gold

Gold is a precious metal. A block of gold 10 cm × 10 cm × 10 cm weighs about 20 kg and would cost £200,000 to buy.

Gold has lots of uses and it is most often used to make jewellery.

Pure gold is very soft. It is mixed with other metals to form gold alloys. The less gold there is in the alloy the harder the alloy is.

Gold alloys can be of two types – yellow gold and white gold.

The composition of gold in an alloy is given by a unit called the carat.

The table gives some information about pure gold and its alloys.

| Alloy | Percentages by weight | | | | | |
	Gold	Silver	Copper	Zinc	Nickel	Palladium
9 carat yellow gold	37.5	10	45	7.5	–	–
9 carat white gold	37.8	0	40	10.4	11.8	–
18 carat yellow gold	75	16	9	–	–	–
18 carat white gold	75	4	4	–	–	17
24 carat pure gold	100	–	–	–	–	–

Questions

1 How many carats is pure gold?

2 Which metals are added to yellow gold alloys?

3 If 12 carat gold was made, what percentage of gold would it contain?

4 Which yellow gold alloy is hardest?

Problem or resource?

In this item you will find out

- the advantages and disadvantages of recycling

- the conditions needed for the rusting of iron

- similarities and differences between iron and aluminium

The photograph shows a car scrapyard. These cars are here because they have come to the end of their useful life. Some may have been in accidents.

Often these cars can provide spare parts for other cars. Otherwise, these cars will be broken down to recover useful materials that can be recycled. Often the car may be crushed so it takes up less space before it is moved.

a Suggest what properties glass needs to have to be used as a car windscreen?

b Steel does not rust easily. Why does this make it suitable for a car body?

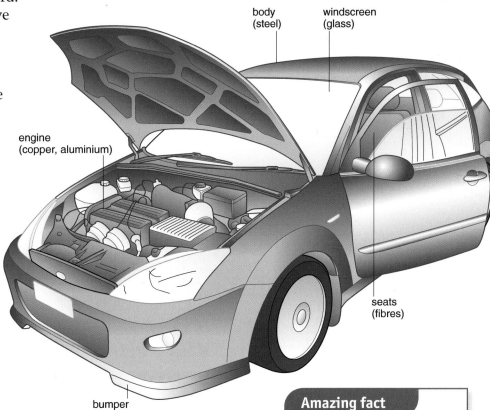

body (steel)

windscreen (glass)

engine (copper, aluminium)

seats (fibres)

bumper (plastic)

▲ *What a car is made from*

Amazing fact

The thickness of steel in a modern car is much less than in a car made 30 years ago. This is because rust-proofing methods have improved and cars do not rust quickly.

▲ *A rusty van*

Recycling cars

Recycling materials is good because it saves natural resources and cuts down on the disposal of waste. However, at the moment the value of old cars for scrap barely covers the cost of collecting them. For that reason cars and vans are sometimes seen abandoned.

In the future car manufacturers will have to modify designs so that more of the materials they use can be recycled to make new cars. When they sell a new car they will have to agree to take it back at the end of its life. By January 2007 they will have to plan to recycle 85% of the car and by 2015 at least 95%.

c **Suggest how these changes could help the environment.**

Rusting

Old cars suffer from **rust**.

The diagram shows an experiment to find what is needed for the rusting of iron to take place.

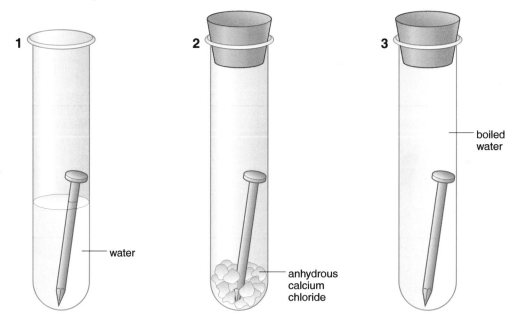

Rusting of iron ▶

Test tube 1 is a control experiment. It shows that an iron nail rusts with air and water. Test tube 2 contains a nail with dry air. The calcium chloride removes any water vapour. No rusting takes place here. Test tube 3 contains a nail in boiled water but with no oxygen or air there, again, no rusting takes place. This experiment shows that rusting takes places when iron is in contact with water and oxygen (air). Rusting is speeded up by salt water or **acid rain**.

In West Africa, the climate is hot and humid. Would you expect rusting to take place faster or slower than in the UK?

Aluminium does not corrode even in moist conditions. The photograph shows an aluminum bucket which has not corroded.

There is no **corrosion** because there is a thin coating of aluminium oxide on the surface of the aluminum. This coating does not flake off and it protects the surface of the aluminium from corrosion.

Iron and aluminium

The table shows some of the similarities and differences between the properties of iron and aluminium.

Property	Iron	Aluminium
density	denser than steel	less dense than iron
magnetism	magnetic	not magnetic
rusting	rusts easily	does not rust easily
malleability	**malleable** (can be bent easily)	malleable (can be bent easily)
electrical conductivity	a good electrical conductor	a good electrical conductor

 People pick lead and copper out of the metal waste from cars. This is shredded into smaller pieces. How can iron be separated from a mixture of iron and aluminium?

As you have seen in Item C2d, alloys are made from two or more metals and they are often more useful than the pure metals. Steel is an alloy containing iron with a very small percentage of carbon. Steel is more useful for making car bodies than iron because:

- it is harder and stronger than iron
- it is less likely to corrode than iron.

Some cars now have bodies made out of aluminium rather than steel. This is more expensive but the car body will not corrode. Also, because aluminium is less dense, the car will be lighter and will therefore use less fuel.

keywords

acid rain • corrosion • malleable • rust

Corrosion of two metals together

There are times when two metals might come in contact in a car. This can sometimes cause problems. A car manufacturer is considering using metal rivets to fix panels of steel together. Either zinc or copper rivets could be used. Angus is carrying out some experiments to investigate what happens when the two metals are in contact. He sets up the following three experiments:

A one with an iron nail
B one with an iron nail in contact with a piece of zinc
C one with an iron nail in contact with a piece of copper.

Angus uses ferroxyl indicator to show where rusting of iron is taking place. Areas that are pink are protected from corrosion. Areas that are green show where corrosion is taking place.

The diagram shows his results.

Reactivity series of metals
Potassium   most reactive
Sodium
Calcium
Magnesium
Aluminium
Carbon
Zinc
Iron
Tin
Lead
Hydrogen
Copper
Silver
Gold
Platinum   least reactive

(elements in italics, though non-metals, have been included for comparison)

▲ Reactivity series

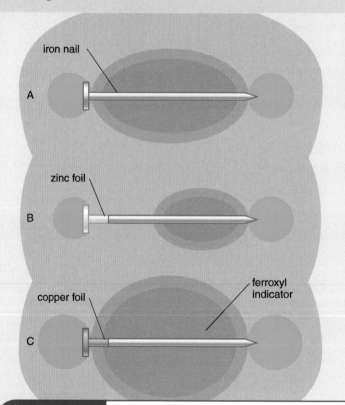

◀ Results of corrosion experiment

iron nail

A

zinc foil

B

copper foil

ferroxyl indicator

C

Questions

1 What can you conclude from the results of the experiments with the three nails?

2 Look at the reactivity series. Write down the three metals (iron, aluminium and copper) in order of decreasing reactivity.

3 Which rivets would you suggest are used? Explain your choice.

4 The manufacturer also makes the car with an aluminum body. He is planning to bolt body panels together. Aluminium bolts are not strong enough. Would you suggest iron bolts? Explain your answer.

Air fit to breathe

In this item you will find out

- the composition of clean air

- how air pollutants are formed and what problems they cause

- how some pollutants are removed

The photograph shows an aerial view of Mexico City.

It is difficult to see any distance because of **atmospheric pollution**. Car exhausts and waste gases from chimneys of houses and factories cause air pollution. As the population grows and industry develops the problem can get out of control. In 1956 there were serious air pollution problems in London and other places in the UK.

The introduction of the Clean Air Act in 1956 greatly reduced emissions. As a result, the air conditions in London are today better than they have ever been.

The World Health Organisation (WHO) has recommendations for air quality. In Lahore in Pakistan the levels of pollutants are 20 times greater than the WHO maximums.

The pie diagram shows the percentage composition by volume of clean, dry air.

Amazing fact

In Lahore, 6.4 million people are admitted to hospital each year with illnesses caused by air pollution. That is about 7% of the city's population.

a Which gas is in the atmosphere in the largest amounts?

b Which gas in the pie diagram is needed for us to breathe?

The other gases in the air include argon and neon. Because air is a mixture, its composition can vary from place to place. But the composition never varies much from that shown.

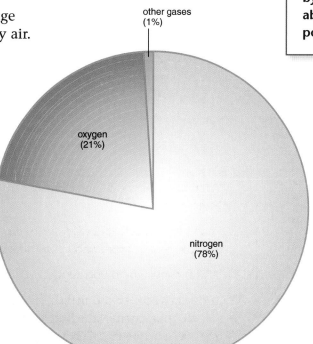

other gases (1%)

oxygen (21%)

nitrogen (78%)

Balancing act

The percentage of carbon dioxide and oxygen in the air remains constant.

The diagram below shows a simple carbon cycle which explains why the percentages remain constant.

The processes of **combustion**, especially of carbon fuels, and **respiration** both use up oxygen and produce carbon dioxide.

Photosynthesis in green plants uses up carbon dioxide and produces oxygen.

The origin of the atmosphere

The composition of the Earth's atmosphere has remained approximately constant for a thousand million years. Before that there were very big changes over the previous four hundred million years.

Originally, the atmosphere existed mainly of gases from inside the Earth. These gases would largely be water vapour, carbon dioxide but with smaller amounts of methane and ammonia.

The steam cooled and condensed to form water.

The ammonia was very slowly converted to nitrogen.

When simple plants were formed carbon dioxide started to be turned into oxygen by the process of photosynthesis.

When simple animals started to exist respiration produced carbon dioxide.

This set up the balance that has kept the composition of the atmosphere constant.

It is, however, easy for this delicate balance to be disturbed, e.g. by destroying rainforests or killing the green algae in the oceans.

C **How is the composition changed if rainforests are destroyed or green algae killed?**

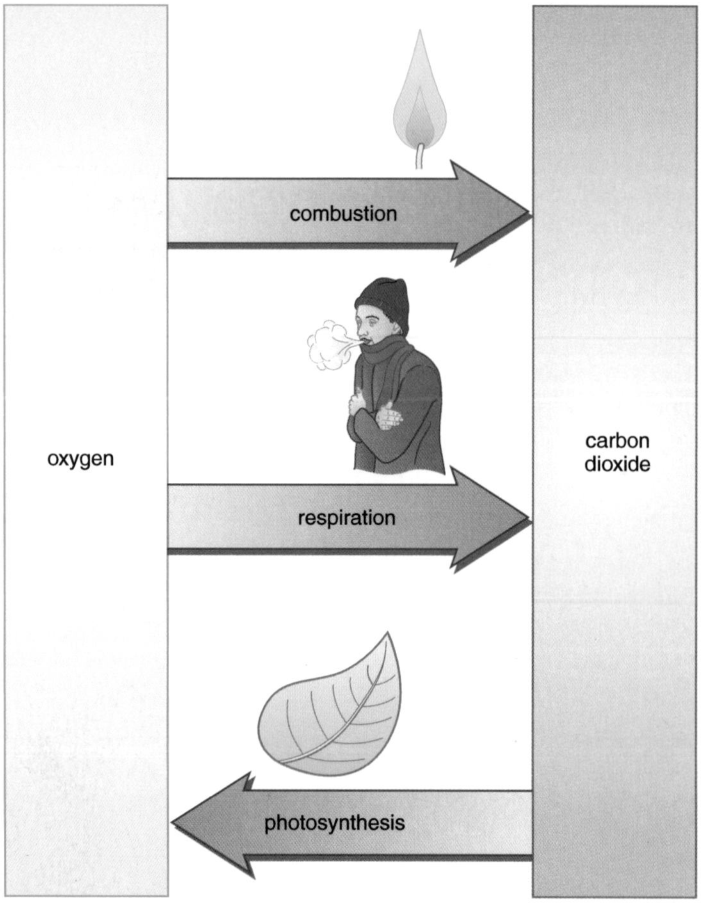

oxygen

combustion

respiration

photosynthesis

carbon dioxide

▲ *Keeping oxygen and carbon dioxide in balance*

Air pollution

The table shows some of the common pollutants in the air, how they are formed and what effects they have.

Pollutant	How it forms	Effects
carbon monoxide	incomplete combustion of petrol or diesel in car engines	poisonous gas
oxides of nitrogen	in car engines from the reaction of nitrogen and oxygen	acid rain and **photochemical smog**
sulfur dioxide	combustion of fossil fuels containing sulfur, for example, coal	acid rain

The photographs above show the effects of acid rain on a stone statue and trees in a forest.

Acid rain will kill plants and pollute water. This kills fish and animals that live in the water. It will also corrode metals and rocks such as limestone.

keywords

• atmospheric pollution • combustion • respiration

Reducing atmospheric pollution

Air pollution has many harmful effects on our health and the environment around us. It is important that air pollution is monitored and controlled. There are several ways that pollution can be reduced. For example, a catalytic converter can be fitted to a car exhaust system to remove the pollutant carbon monoxide. It converts carbon monoxide in the exhaust gases to carbon dioxide.

▶ *In the UK, any car with too many harmful emissions fails its MOT test and cannot be used*

Asthma and atmospheric pollution

Health Protection Agency researchers examined the health of people living near a factory producing iron in Birmingham. In 1997, the company which operates the factory had introduced a programme to reduce emissions. So, researchers examined the impact this had on hospital admissions from asthma in the area.

Here are their results.

Years considered	Number of patients admitted to hospital with asthma
1995–1997	1 200
1998–2000	840

The researchers did not examine whether many people moved in or out of the area during this time, which could affect hospital admission rates. Also, they did not look at whether the high admission rates before the factory cut its emissions could have been caused by something else.

Their data has added to the debate over whether there is a link between asthma and air pollution. Some studies have suggested there is a link while others have not. For example, asthma rates have soared in Britain since the introduction of the Clean Air Act in 1956, which cut atmospheric pollution by a large amount across the country.

Questions

1 There was a decrease in the number of patients admitted. Calculate the percentage decrease.

2 What did the research appear to show?

3 Why was this research not truly scientific?

4 There seems to be a link with increasing air pollution and the number of cases of asthma. What evidence contradicts this?

Closer and hotter

In this item you will find out

- that reactions take place at different rates
- that a reaction occurs when particles collide
- that increasing temperature or concentration speeds up the reaction

Here are four photographs of chemical reactions taking place.

▲ An explosion

▲ A tablet fizzing in water

▲ A gravestone eroding

The four reactions are taking place with different **rates of reaction**.

The reaction with the fastest rate will be over quickly. The one with the slowest rate will take the longest time.

 Arrange the four reactions in order. Start with the slowest and finish with the fastest.

A chemical reaction takes place when particles of the **reactants** collide with each other. The more collisions that take place the faster will be the reaction.

You can speed up a chemical reaction in three ways:

1 by increasing the temperature
2 by increasing the **concentration**
3 by increasing the pressure (if the reactants are gases).

▲ A car rusting

135

Increasing reaction rates

If you increase the temperature, the particles move faster so they have more energy. They collide more often and this increases the rate of the reaction. As shown in the diagram on the left.

If you decrease the temperature then the rate of reaction slows down.

If you increase the concentration (or the pressure of a gas reaction), then the particles are more crowded together. This also increases the rate of reaction.

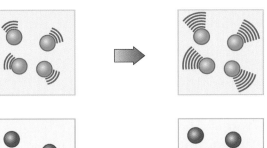

If the acid is heated the particles move faster.
More collisions occur as temperature rises.

▶ *More particles crowded together mean more collisions*

If the concentration of the acid is increased, there are more acid particles in the same volume of water. More collisions occur when more particles are crowded together.

Measuring the rate of a reaction

You can carry out an experiment to investigate the rate of reaction between two reactants.

The diagram below shows apparatus that can be used to investigate the rate of reaction between marble chippings and acid.

We are going to react marble chippings (calcium carbonate) and dilute hydrochloric acid together. The word equation shows what is produced in the reaction. The amount of **product** formed depends on the amount of reactants used.

calcium carbonate + hydrochloric acid → calcium chloride + water + carbon dioxide

> **Examiner's tip**
>
> **Remember this experiment works because one of the products, carbon dioxide, is a gas. The volume is measured at intervals.**

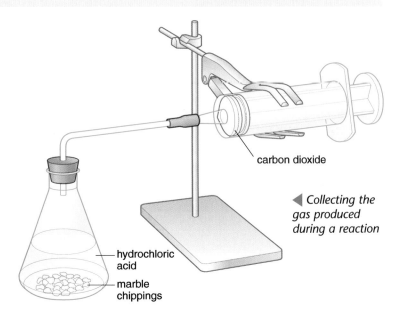

carbon dioxide

◀ *Collecting the gas produced during a reaction*

hydrochloric acid

marble chippings

The graph shows the results.

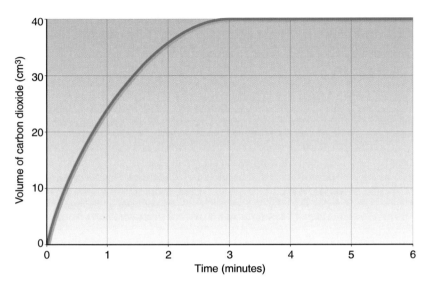

▲ *Graph of the reaction between marble chippings and acid*

As you can see, the graph is a curve. It starts steeply. As the reaction proceeds the gradient of the graph reduces and reduces. Finally, when the reaction has stopped the graph is horizontal. The reaction stops when one of the reactants is used up.

b **After what time does the reaction stop?**

At the end of the experiment there is no calcium carbonate in the flask. This means that the calcium carbonate has been used up.

c **How many cm^3 of carbon dioxide form between 0 and 1 minute?**

d **How many cm^3 of carbon dioxide form between 1 and 2 minutes?**

You will notice that when the graph is steep a large volume of gas is produced and the reaction is fast.

The reaction is repeated. The hydrochloric acid is mixed with an equal volume of water. The concentration of hydrochloric acid is half what it was in the first experiment.

Here are the results.

Time (minutes)	0	1	2	3	4	5	6
Volume of gas (cm^3)	0	14	24	32	37	39	40

Now the concentration is less the reaction takes longer. It is a slower reaction. If you plot this as a graph you will see that it is less steep and takes longer to go horizontal. At the end of the experiment the same volume of carbon dioxide is produced because the same mass of calcium carbonate was used in the experiment.

keywords

concentration • product • rate of reaction • reactant

Investigating temperature

Dr Sarah Weston is carrying out some rate of reaction experiments to see how changing temperature affects the rate of reaction.

She is collecting and measuring the volume of a gas produced, at intervals.

She carries out two reactions, A and B.

The first reaction is carried out at 60°C.

The second at 90°C.

Dr Weston uses the same amount of reactants in each experiment.

She plots the two graphs on one grid.

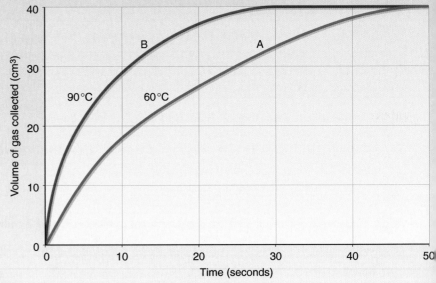

Questions

1 Choose from the list the best piece of apparatus to measure the gas given off in this experiment.

   boiling tube   50 cm³ gas syringe   100 cm³ gas syringe

   100 cm³ measuring cylinder   pipette

2 After how many seconds is reaction A completed.

3 Which reaction has the fastest rate at the start? How do you know this from the graph?

4 Sketch a graph to show what you think would happen if the reaction was repeated at 120°C.

5 What happens to the amount of product formed when a higher temperature is used?

Explosions and catalysts

In this item you will find out

- about explosions

- how catalysts speed up chemical reactions

- how increasing the surface area of a reactant speeds up a reaction

We all know that some substances can explode and substances such as TNT or dynamite are very dangerous.

Look at the photograph below of flour, icing sugar and custard powder. Do you know these are dangerous explosives?

Manufacturers using these materials have to take special care when handling these materials as they can cause **explosions**. Special ventilation is needed in the factory to remove the dust.

An explosion is a very fast reaction that releases a large volume of gases.

▲ Flour caused this explosion

▲ Flour, icing sugar and custard powder

Have you ever tried to light a coal fire? It is not an easy thing to do. Coal does not catch alight easily. Now, coal dust is a different matter. About 60 years ago there was a big explosion underground in a coal mine in Stoke-on-Trent that killed 51 men. A mixture of coal dust and air caused it.

You are used to making a small explosion in the laboratory when you test for hydrogen. Remember when you put a lighted splint into hydrogen it burns with a squeaky pop. Have you noticed you get the best pop when hydrogen and air are mixed?

Amazing fact

The Great Fire of London, in 1666, started in a baker's shop. No-one knows what sparked the fire but flour dust probably made it worse.

Catalysts

A **catalyst** is a substance which changes the rate of a chemical reaction. At the end of the reaction the catalyst is unchanged. Because of this, a small amount of catalyst will speed up large amounts of reactants. Catalysts are widely used in industry to speed up chemical reactions. The reactions never produce more than would be produced without a catalyst but they produce it more quickly.

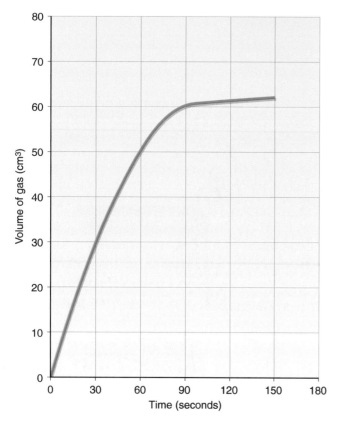

Hydrogen peroxide is widely used as a bleach. Some people use it to bleach hair. It is compound of hydrogen and oxygen. It decomposes very, very slowly to form water and oxygen.

hydrogen peroxide → water + oxygen

Manganese(IV) oxide is a catalyst for this reaction. The photographs show some hydrogen peroxide in a beaker before and after adding powdered manganese(IV) oxide.

The manganese(IV) oxide is not used up and remains unchanged at the end of the reaction. The graph shows the decomposition of $25\,cm^3$ of hydrogen peroxide with $0.2\,g$ of manganese(IV) oxide

a Look at the graph and read the volume of gas collected after 30 seconds.

b What mass of manganese(IV) oxide is left at the end of the reaction?

c What volume of gas is produced between 30 and 60 seconds?

d How can you tell from the graph that the reaction is faster after 30 seconds than it is after 60 seconds?

Increasing surface area

The rate of a reaction can also be increased by using a powdered reactant instead of a solid chunk. The reaction is faster because the powder has a larger **surface area** and so there are more particle collisions. This makes the reaction happen more quickly. This is shown in the diagram.

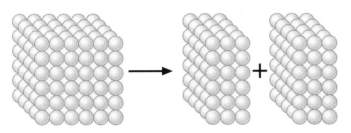

▲ *Splitting a lump gives a larger surface area*

The surface area is bigger because more area is exposed.

The particles collide more frequently, and so the reaction gets faster.

For example, powdered calcium carbonate reacts much faster than chippings of marble. The graph shows the results of experiments with chippings of marble and powdered calcium carbonate.

 Is this reaction faster when powder is used? How can you tell from the graph?

The photographs show the difference between the reaction with a chunk and with the powder.

keywords

catalyst • explosion • surface area

Epoxy adhesives

Epoxy adhesives are used very widely today. They were developed during the Second World War to stick panels of aircraft together. This was easier than using rivets or bolts.

The adhesive comes in two separate tubes. In one tube is the resin paste and in the other tube is the hardener.

Before you can use them, they have to be mixed. The hardener is a catalyst.

After the hardener is mixed with the resin, the mixture sets hard. Sometimes the hardening can be speeded up by heating the adhesive.

The table shows the hardening time of different mixtures of resin and hardener.

Percentage of resin	70	75	80	85	90	95
Percentage of hardener	30	25	20	15	10	5
Hardening time (minutes)	5.0	5.0	5.0	7.0	9.0	12.0

A related product is used to make car body fillers. Again, the resin and the hardener are mixed. They quickly set to fill a dent in the car body. After drying they can be sanded and then painted.

Questions

1 After mixing the resin and the hardener, the mixture cannot be stored. Why is this?

2 Only a small amount of hardener is needed. Suggest why.

3 How can the hardening be speeded up?

4 The hardener is much more expensive than the resin. Use the data in the table to explain the best mixture of resin and hardener to use to mend a broken ornament, quickly and cheaply.

C2a

1 Use words from the list to finish the sentences.

colloid   phosphorescent   pigment
solution   solvent   thermochromic

The colour of a paint is due to the type of ____(1) used.
In a water-based paint the water acts as the ____(2).
A paint is a mixture called a ____(3).
A paint that glows in the dark is called ____(4).
A paint that changes colour when heated or cooled is ____(5).   [5]

2 Match the words on the left with the correct definitions on the right.

1 extender   a thins the paint
2 solvent   b increases the bulk of the paint
3 binding medium   c gives the paint its colour
4 pigment   d forms a hard paint film   [3]

3 Years ago, gloss paints had to be stirred before use and, from time to time, during use.

It is not necessary today. Suggest why.   [2]

4 For many centuries cloths have been dyed using plant dyes such as woad. Now synthetic dyes are used. Suggest two reasons why synthetic dyes have replaced natural dyes.   [2]

5 A wood varnish consists of a natural resin dissolved in an organic solvent.
Describe the change that takes place when the varnish dries.   [1]

C2b

1 Limestone and marble are forms of:
A calcium carbonate   B calcium oxide
C sodium carbonate   D sodium oxide   [1]

2 Splitting up limestone by heating is an example of:
A combustion   B neutralisation
C precipitation   D thermal decomposition [1]

3 Choose three materials from the list that are used to make concrete.
cement   gravel   lime   soda   water   [2]

4 Write down two problems caused when rocks are quarried.   [2]

5 Write down the name of a construction material made from:
  a bauxite (aluminum ore)
  b clay
  c sand and sodium carbonate   [3]

6 Put these three rocks in order of increasing hardness:
granite   limestone   marble   [2]

7 Which two materials heated together produce cement?   [2]

8 Use chemicals from the list to write the word equation for the thermal decomposition of limestone.
calcium carbonate   calcium oxide
carbon dioxide   water   [2]

C2c

1 Which two of the following occur when plates move?
earthquakes   tides   rain storms
tsunamis   [2]

2 Answer the questions which follow using words from the list.
core   crust   lithosphere   mantle
tectonic plate

  a What is the name given to the outer 1000 km of the Earth?
  b What is the name of the molten central part of the Earth?
  c What is the name of one of the large sections of the crust that move slowly?
  d What is the name of the partially fluid rock below the crust?   [4]

3 Write down one way molten rock can get onto the Earth's surface.   [1]

4 **a** What causes an earthquake?
  b Where on the Earth are earthquakes most likely to happen?   [2]

5 Explain why some igneous rocks are made up of small crystals and others large crystals.   [2]

C2d

1 Which two of the metals are alloys?
A aluminium   B brass
C copper   D steel   [2]

2 **a** Describe some of the problems of recycling copper.   [2]
  b Why is it particularly important to recycle copper?   [1]

3 Duralumin is an alloy made from aluminium and copper. It is denser than pure aluminium but stronger than pure aluminium or pure copper.

   a Despite being denser than pure aluminium, duralumin is used for building aircraft rather than pure aluminium. Why is this? [1]

   b Overhead power cables are made from pure aluminum rather than duralumin. Suggest why. [1]

   c Pure aluminium costs £800 per tonne and pure copper costs £1200 per tonne. Work out the cost of the metals used to produce 1 tonne of duralumin containing 10% copper. [2]

C2e

1 Match the metal on the left with a place in a car where the metal is most often used.

   1 aluminum   a car battery
   2 steel   b electrical wiring
   3 lead   c alloy wheels
   4 copper   d car body   [3]

2 Which word is used for a mixture of two or more metals?
   A alloy   B compound   C element   [1]

3 Finish the sentences using words from the list.
dense   magnetic   malleable   shiny

   Iron and aluminium can be made into thin sheets. This is because they are ____(1).
   A car made of steel is heavier than the same car made of aluminum because steel is ____(2).
   Iron and aluminum can be separated because iron is ____(3).   [3]

4 Many plastics and fibres in cars are made from products from crude oil. Suggest why recycling these materials is important for the environment. [2]

5 Suggest reasons why steel is used instead of iron for car bodies. [3]

C2f

1 Answer the questions using words from this list.
carbon dioxide   hydrogen   nitrogen oxygen   water vapour

   a Which two substances are not found in dry air? [2]
   b Which gas increases in concentration when fossil fuels burn? [1]
   c Which gas increases in concentration when photosynthesis takes place? [1]

2 Match the pollutant on the left to a problem the pollutant causes on the right.

   1 CFCs   a poisonous gas in car exhausts
   2 sulfur dioxide   b depletes ozone layer
   3 nitrogen dioxide c causes acid rain but not photochemical smog
   4 carbon monoxide d causes acid rain and photochemical smog   [3]

3 Which substance can remove sulfur dioxide from the waste gases of a fossil fuel power station?
   A powdered limestone   B sodium hydroxide
   C sulfuric acid   D water   [1]

4 Look at the graph showing the number of deaths each day in London between 1 December 1952 and 15 December. It also shows the concentrations of smoke and sulfur dioxide.

   a How many deaths occurred on 5 December 1952? [1]
   b What is the relationship between the number of deaths and the concentration of sulfur dioxide? [2]

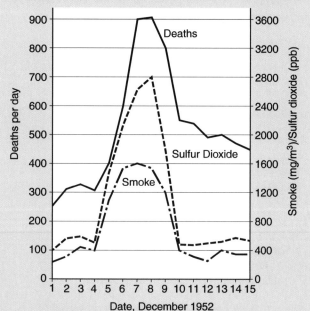

5 The original atmosphere of the Earth included hydrogen and helium. Why did they escape from the atmosphere? [1]

6 How are nitrogen oxides formed in a car engine? [1]

7 The diagram summarises an experiment using iron filings. Explain the results of the experiment.

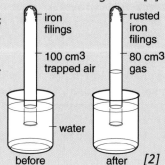

[2]

C2g

1 Strips of magnesium ribbon were added to four solutions of hydrochloric acid of different concentrations. The time was measured until all the magnesium disappeared. All experiments were carried out at 20 °C.

The results are shown in the table.

Sample	A	B	C	D
time for magnesium to react (seconds)	35	15	12	25

   a Which experiment was fastest? [1]
   b Arrange the four solutions in order of increasing hydrochloric acid concentration. [3]
   c Experiment with sample D was repeated at 30 °C. Which is the most likely result in seconds?

   17   25   28   35 [1]

2 The graph shows the volume of hydrogen collected at intervals in an experiment with dilute sulfuric acid.

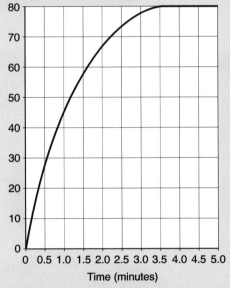

   a After how long is the reaction complete. [1]
   b After how long are half the reactants used up? [1]
   c Calculate the average volume of gas produced each second in the first minute. [2]
   d Why is your answer to **c** not a useful measure of the rate of reaction? [1]

C2h

1 A catalyst is a substance which:

   A will start a reaction   B speeds up a reaction
   C stops a reaction   D evolves more energy [1]

2 The mass of a catalyst was determined at intervals during a reaction. Which one of the graphs A–D would be obtained?

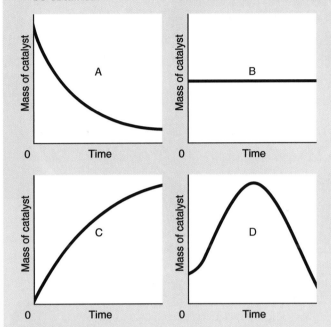

3 An experiment was carried out using marble chips and dilute hydrochloric acid, to investigate the effects of particle size on the rate of reaction.

$$CaCO_3 + 2HCl \rightarrow CaCl_2 + CO_2 + H_2O$$

A large marble chip was placed in a conical flask and the flask placed on a top pan balance.

25 cm³ of hydrochloric acid was added to the flask and a plug of cotton wool was placed in the neck of the flask. The reading on the balance was noted at intervals.

The results are shown in the table.

Time (min)	0	2	4	6	8	10	11	12
Total loss in mass (g)	0.0	2.2	2.9	3.3	3.6	3.7	3.7	3.7

   a Why is there a loss of mass during the experiment? [1]
   b Plot a graph of the total loss of mass (on the y-axis) against time. [3]
   c When was the reaction fastest? [1]
   d After how many minutes was the reaction completed? [1]
   e After 12 minutes a small piece of marble remained in the flask. What can be concluded from this observation? [1]

P1 Energy for the home

My electricity bill is too big. What can I do to save money?

My new laptop connects to the Internet through a wireless network. How does the link work?

Why do I have to put on sun cream when I go to the beach?

- In the UK, we use more energy in our homes to keep them warm than all of the other uses put together. Using that energy carefully can save a lot of money. The waves that we use for heating our food also allow information to be exchanged over the Internet.

- Keeping our houses warm and comfortable doesn't have to use a lot of expensive energy. We just have to build our houses from the right materials and put energy saving as a high priority in their design.

What you need to know

- Energy can be transferred from one form to another.

- Heat energy passes readily through conductors and slowly through insulators.

- Solids, liquids and gases are made of particles.

- Light is a wave, travels in straight lines and can be reflected.

Warming up

In this item you will find out

- the difference between heat and temperature

- what makes heat energy move from one place to another

- about specific heat capacity and specific latent heat

Coffee made from cold tap water is disgusting. The **temperature** of the water is only about 10 °C, not enough to make the instant coffee granules dissolve properly.

To get the water to 100 °C, you need to add **heat** to it. So you put it in a kettle. What's the quickest way of making coffee? Easy – by heating just enough water to fill the mugs.

Suppose you put too much water in the kettle? It takes longer to boil … but why? The water needs more energy, and the kettle can only supply so much.

a   **Why does a full kettle take longer to boil than one which is nearly empty?**

b   **How is temperature different from heat?**

Most electric kettles switch themselves off when the water starts to boil. Will your coffee be hotter if the water boils for longer?

Surprisingly, it won't. Water boils at 100 °C, no matter how much heat you add to it!

Heat and temperature are not the same things. Temperature is a measurement of how hot or cold an object is, while heat is a measurement of **energy**.

Temperature is measured in °C (degrees Celsius) and heat energy is measured in J (joules).

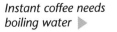

Instant coffee needs boiling water ▶

Energy flow

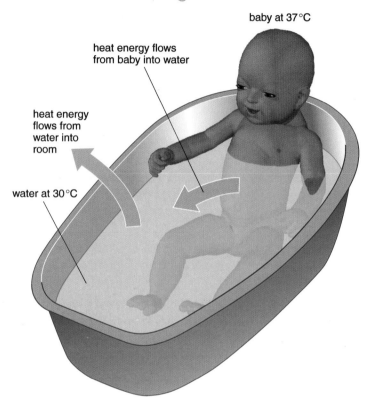

room at 20°C

baby at 37°C

heat energy flows from baby into water

heat energy flows from water into room

water at 30°C

▲ Heat energy flows from the baby into the bath water and then into the room

What happens to a hot cup of coffee if you leave it for too long without drinking it? The coffee cools down. The coffee has a higher temperature than the rest of the room, so it loses heat. The hotter the coffee is, the quicker it will cool down. The opposite happens with a cold ice cube. If you leave an ice cube at room temperature it will melt. This is because it has a lower temperature than the room, so it gains heat from the room.

Babies are not good at controlling their temperature. So it is important that their bath water is neither too hot nor too cold. The ideal temperature for the water is 37°C – the same temperature as the human body.

Heat energy always flows from a hot object to a cooler one. This causes hot objects to cool down and cool objects to warm up. If the water is only 30°C, the baby will get cold because heat energy will flow from the baby into the water.

c Suppose the bath water is 39°C. Which way will the heat energy flow?

Specific heat capacity

The **specific heat capacity** (shc) of a material tells you how much energy, in joules, you need to raise the temperature of 1 kg of the material by 1°C. It is the amount of energy the material can hold. Specific heat capacity is measured in units of joules per kilogram per degree celsius (J/kg/°C). It is different for different materials because some materials need more energy than others for the same temperature change.

The table tells you that sand needs twice as much energy as brass to get the same temperature change.

Material	Specific heat capacity (J/kg/°C)
brass	400
steel	500
glass	700
sand	800
wax	2 900

d Which material needs the most energy to raise its temperature? Which material needs the least energy?

The amount of energy that you need to put into a material to change its temperature depends on three things:

• the mass of the material
• the type of material
• the temperature change you want to achieve.

From ice to steam

If you want to make steam from a solid block of ice at 0°C then you need to add a lot of energy.

Temperature (°C) (vertical axis)

100 — boiling point of water

water boils into steam

water heats up

ice melts

0 — melting point of ice

ice heats up

Time of heating (horizontal axis)

thermometer

beaker

ice

water

heater

▲ Heating a block of ice turns it into water then steam

You need to add energy to the ice to melt it – to make it into a liquid at 0°C. Each kilogram of ice needs 330 000 J for this. You can see from the graph that while the ice is melting, energy is being added, but the temperature of the water does not rise. Then you need more energy to increase the temperature of the water to 100°C, another 420 000 J for each kilogram. Finally, you need another 2 300 000 J to boil a kilogram of water into steam. Again, you will see from the graph that while the water is boiling, energy is being added, but the temperature of the water does not go above 100°C.

Specific latent heat

Sometimes a cold drink is better than a hot one. Ice cubes are the answer!

When you add ice to a glass of water, heat energy flows from the water at 20°C to the ice at 0°C and the temperature of the water drops. This energy is needed to change the solid ice into liquid water and is called latent heat.

The **specific latent heat** of a material tells you how much energy is needed to boil or melt 1 kg of that material. It is different for different materials. If you want to melt a 1 kg block of ice then you need to add 330 000 J of energy. So the specific latent heat of ice is 330 000 J/kg.

e Show that you need 165 000 J to melt 0.5 kg of ice.

Amazing fact

Melting an ice cube in a glass of water can lower the temperature of the water by 10°C.

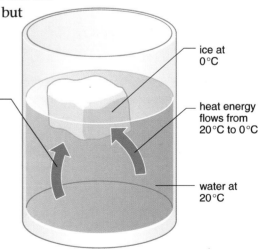

heat energy from the water melts the ice

ice at 0°C

heat energy flows from 20°C to 0°C

water at 20°C

▲ Heat energy from the water melts the ice

keywords

energy • heat • specific heat capacity • specific latent heat • temperature

Keeping cool

Ailsa and Jon rely on their fridge. They do a weekly shop, put a lot of it in the fridge and take it out it as and when they want to eat it. The fridge allows them to keep food fresh for longer. So how does it work?

The key material is the refrigerant, which is a liquid that boils at about 0°C. As the refrigerant boils in pipes inside the fridge, it takes in heat energy from the food in the fridge and turns from a liquid to a gas. This keeps the fridge cold. The gas is then compressed by a pump, forcing it to become a liquid once more, but this time in pipes outside the refrigerator. For a gas to condense into a liquid, it has to lose energy. The heat energy from the gas in the pipes is transferred to the kitchen which heats it up a bit.

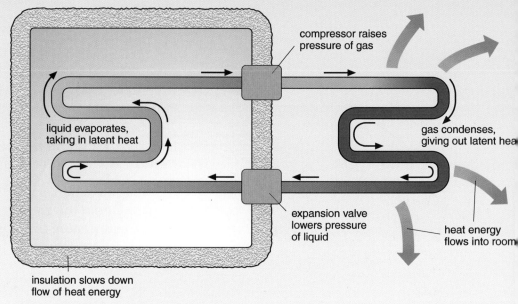

▲ How a fridge works

Ailsa and Jon keep an eye on the temperature of their fridge. Too cold and it will freeze the food. Too warm and the food will not stay fresh for long. The fridge uses an electronic thermometer to display its temperature with a row of lights.

Questions

1 What is the ideal boiling point of a refrigerant?

2 Explain why the boiling refrigerant cools the inside of the fridge.

3 What does the pump do to the refrigerant?

4 Explain why the fridge heats up the kitchen.

5 Why do Ailsa and Jon need to check the temperature of their fridge?

Keeping heat in

In this item you will find out

- how to reduce energy loss by insulating a house

- how to compare the cost of the insulation with the fuel savings

- about energy efficiency

In the UK, the average outside temperature is about 5 °C in winter. So we stay indoors, where we can heat our houses to around 20 °C.

If our homes have poor **insulation** then much of the heat energy escapes to the outside. It costs a lot to replace the heat energy as it flows from the hot inside to the cold outside.

Look at the thermogram on the right. It shows the heat energy being lost at the surface of the house.

Each colour shows a different temperature – red for hot, blue for cold. Hot surfaces lose heat energy more rapidly than cold ones.

a **Which part of the house is losing lots of heat?**

b **If the heat energy that escapes from the house is not replaced, what will happen to the temperature of the house?**

The roof is blue. It is well insulated with loft insulation to slow down the flow of heat energy. This cuts down on the cost of heating the house. But insulating the loft space is expensive. Can it be paid for by the money saved on the heating bills?

Amazing fact

It costs about £10 billion to heat all the houses in the UK.

Holding in heat

There are many ways of insulating a house. They all cut down on heat loss, but some are more cost-effective than others.

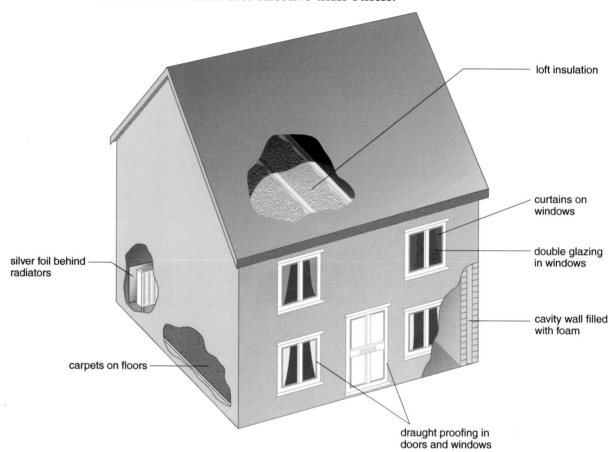

- loft insulation
- curtains on windows
- double glazing in windows
- cavity wall filled with foam
- silver foil behind radiators
- carpets on floors
- draught proofing in doors and windows

c **How many different ways of insulating a house can you see?**

A material which is a good **conductor** of heat will help a house to lose heat energy but a material which is a bad conductor of heat acts as an insulator. Air is a bad conductor of heat and so it is used in lots of insulation materials.

Replacing heat

How do you keep a house at a steady comfortable temperature? You have to replace every joule of heat energy which escapes! There are several ways of doing this, each with its own installation and running costs.

Method of heating	Installation cost	Running cost
portable electrical heaters	cheap	expensive
electrical storage heaters	moderate	cheap
gas or oil central heating	expensive	cheap
wood or coal fires	expensive	expensive

d **State one good feature and one bad feature for each method of heating.**

Counting the cost

Here are some facts about insulating a typical house.

Method of insulation	Installation cost	Fuel saving	Payback time
radiator foil	£5	£10 per year	0.5 year
loft insulation	£200	£100 per year	2 years
cavity wall insulation	£900	£150 per year	6 years
double glazing	£2 000	£50 per year	40 years

Think about draught-proofing for the doors and windows to save heat energy:

• the draught-proofing will cost about £80
• it will save about £20 per year in heating costs.

The **payback time** tells you how cost effective the draught-proofing will be.

$$\text{payback time} = \frac{\text{installation cost}}{\text{fuel saving}}$$

$$\text{payback time} = \frac{£80}{£20 \text{ per year}} = 4 \text{ years}$$

This is better than double glazing, but worse than loft insulation!

 Building a porch onto a house costs £1000, but saves about £50 in heating costs. Calculate the payback time.

porch stops cold air entering house when people enter or leave

▲ *Adding a porch cuts down on draughts and saves money*

keywords

conductor • insulation • efficiency • payback time • useful energy output • total energy output

Energy efficiency

The **efficiency** of a heating system tells you how much of the energy you put into the system goes into the house as useful heat energy. What doesn't become heat energy may be wasted as light or sound. Here are some examples.

Heating system	Total energy output	Useful energy output	Efficiency
electrical heater	100 J	100 J	1.0
gas central heating	100 J	80 J	0.8
wood fire	100 J	40 J	0.4

Efficiency can be calculated using this equation:

$$\text{efficiency} = \frac{\text{useful energy output}}{\text{total energy output}}$$

f **For every 100 J of electrical energy that is put into a fan heater, only 95 J comes out as heat energy. Calculate its efficiency.**

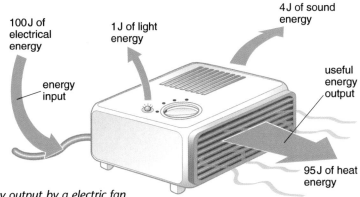

100 J of electrical energy

energy input

1 J of light energy

4 J of sound energy

useful energy output

95 J of heat energy

Most of the energy output by a electric fan heater becomes useful heat energy ▶

Heating their house

Giles and Cherry have just bought their first house. They can't afford much, so they've had to buy an old one. It's so old that all of the rooms have fireplaces which burn coal.

'I don't like coal' says Giles 'It's dirty and expensive. Besides, burning fossil fuels is bad for global warming. We'll need to heat the house some other way.'

Cherry likes the idea of central heating throughout the house, but she knows that they can't afford it. So they have to go for electrical heaters in each room, as the cheapest option.

But when they get their first electricity bill, they have second thoughts – they need to buy a lot of electricity to keep their house warm.

Then Giles meets Jim at work, and Jim, gives him the idea of electric storage heaters. These work by heating up blocks of concrete during the night when electricity is cheap. Fans then blow air through the blocks during the day to transfer the heat energy into the house.

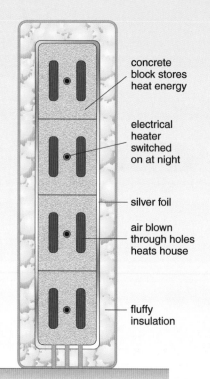

concrete block stores heat energy

electrical heater switched on at night

silver foil

air blown through holes heats house

fluffy insulation

▲ How a storage heater works

'The real problem is the size of the heater.' says Jim. 'All the insulation around the blocks makes the heater quite large.

However, the great thing is that they pay for themselves after only a few years.'

Questions

1 Explain how a storage heater works.

2 Why are the blocks surrounded by a lot of insulation?

3 Each storage heater costs £50, and saves £10 of electricity each year. How long is the payback time?

4 Suggest the disadvantages of storage heaters compared to normal electrical heaters.

Heat energy gets around

In this item you will find out

- about conduction, convection and radiation

- why many insulators contain trapped air

Have you ever wondered why most of your hair is on your head? It is to keep heat energy in your skull!

Heat energy is transferred from chemical energy in your brain to the blood. A head of hair allows most of that heat energy to be carried somewhere useful rather than disappearing into your surroundings.

Your body uses other mechanisms to keep you at a steady temperature of 37°C:

- when you get too hot, extra heat energy radiates from your skin
- layers of fat under your skin slow down the **conduction** of heat energy away from your body
- the blood circulating around your body convects heat energy, sharing it evenly.

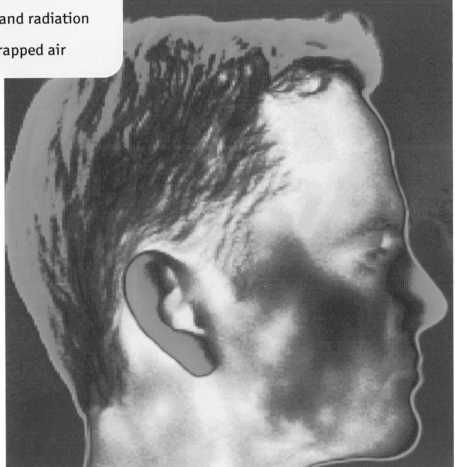

▲ This thermogram shows that heat is lost from your cheeks but held in by your hair

Three mechanisms allow heat energy to flow from hot areas to cold areas, but each one is completely different:

- **radiation** uses waves to carry the heat energy, even through empty space
- conduction passes kinetic energy from one molecule to another, in solids or liquids, through vibrations
- **convection** passes heat energy from one place to another through the flow of liquids or gases.

a State the three ways in which heat energy moves around.

b For each way in which heat energy moves around, give an example in your body.

155

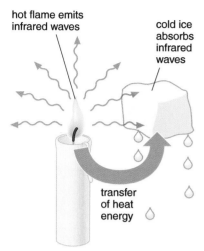

hot flame emits infrared waves

cold ice absorbs infrared waves

transfer of heat energy

▲ Infrared radiation emitted by the hot flame is absorbed by the cold ice

Infrared radiation

Everything loses heat energy through waves called **infrared** radiation. The waves are similar to light, but with a different wavelength so that they are invisible. You can feel them, but you can't see them.

Hot objects emit much more infrared radiation than cold ones. So a hot object placed in a cool place radiates heat energy until they both reach the same temperature.

Two properties of infrared radiation are used to control the heat energy flow from houses:

• white or shiny surfaces don't absorb infrared, they reflect it
• white surfaces emit a lot less infrared than coloured ones.

c Why does a layer of shiny foil between a heater and the wall save money?

d Why do you think lowering the temperature of a central heating thermostat in a house saves money?

Convection

Hot air always rises above cooler air, carrying heat energy away from the roof and walls of a house. The hot air is replaced by falling cold air. Hot water flowing in central heating pipes carries heat energy from a heater to the radiators. Both of these are examples of convection, where heat energy is carried around by moving liquids or gases.

e Explain how your clothes reduce heat loss by convection.

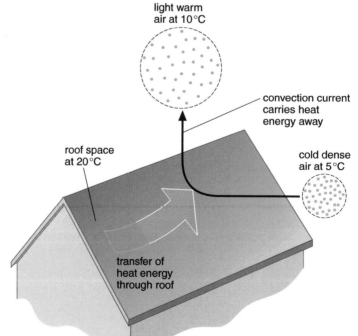

light warm air at 10°C

◀ Warmed air floats above cold air, carrying heat energy away from the roof by convection

convection current carries heat energy away

roof space at 20°C

cold dense air at 5°C

transfer of heat energy through roof

Conduction

Heat energy can flow through solid walls. How is this possible?

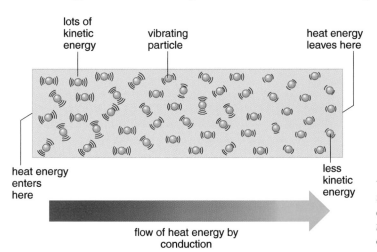

lots of kinetic energy

vibrating particle

heat energy leaves here

heat energy enters here

less kinetic energy

flow of heat energy by conduction

◀ *The vibrations of the particles carry heat energy by conduction from the hot end to the cold end*

The particles which make up solids, liquids and gases never stop moving. The kinetic energy of this endless random motion is what we call heat energy. In solids, where particles are stuck in place, their heat energy makes the particles vibrate.

 What is the difference between conduction and convection?

House insulation

Insulating materials in houses are full of tiny holes. This slows down heat flow through them in two ways:

• heat energy has to conduct through the thin solid between the holes
• the air in the holes is trapped, so the heat energy cannot be transferred by convection.

▶ *This house is being built with lots of wall insulation*

 Cavity walls in houses are often filled with plastic foam. How does this help keep the house warm?

keywords

conduction • convection • infrared radiation

Moon base

NASA is planning to send astronauts back to the Moon in the next few years. They want to set up a Moon base, where astronauts can live for months at a time. One problem is the temperature inside the Moon base. It needs to be kept at a comfortable +20°C all the time.

So what is the problem? There should be plenty of sunshine during the day, as there is no atmosphere or clouds to get in the way of heat energy fron the Sun. But at night (which lasts two weeks) the base will be exposed to outer space at –269°C, about as cold as it is possible to be. It will not be possible to completely stop heat energy flowing from the living quarters at +20°C to space at –269°C, but there are lots of things that can slow it down.

Any heat lost from the surface will have to be replaced. Using fuel transported from Earth is an expensive option, so solar panels and batteries are an alternative. Building the Moon base underground will help to cut down on radiation loss from the surface. A clever choice of materials for the walls and floor should cut down on conduction loss. There is no worry about convection loss as there is no atmosphere on the Moon!

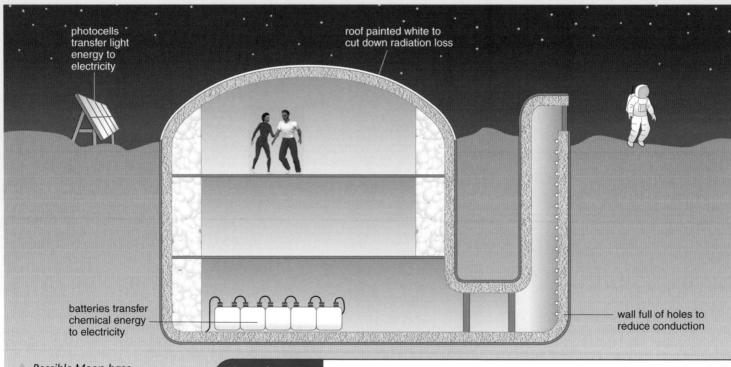

photocells transfer light energy to electricity

roof painted white to cut down radiation loss

batteries transfer chemical energy to electricity

wall full of holes to reduce conduction

▲ Possible Moon base

Questions

1 Explain why the Moon base can't lose heat energy by convection.

2 Explain why the white shiny surface of the Moon base cuts down on heat loss.

3 The outer wall of the Moon base could be made from a metal foam with lots of holes in it. Suggest why.

4 How do the solar panels and batteries provide heat for the living quarters?

5 Suggest why the stores of fuel and food and water are placed between the living quarters and the outer wall.

Food and phones

In this item you will find out

- how infrared and microwaves can be used to cook food

- how microwaves can be used to transfer information between mobile phones

- why some people are worried about the use of mobile phones

Have you ever worried about what your mobile phone might be doing to your brain? After all, it emits **microwaves**, the same waves which cook food in a microwave oven! As the waves from your phone pass through your brain, they heat it up, but only by a very tiny amount. It should not be enough to cause any damage at all, but not everyone is convinced.

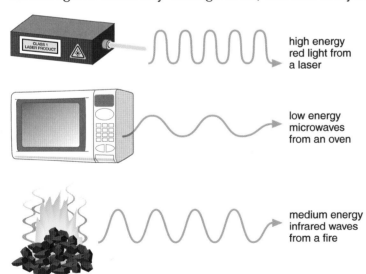

high energy
red light from
a laser

low energy
microwaves
from an oven

medium energy
infrared waves
from a fire

◀ *Microwaves, infrared and visible light are all part of the electromagnetic spectrum*

Food can also be cooked in a conventional oven, where it is heated by infrared radiation emitted by hot surfaces. Although it is not obvious, both infrared and microwaves are similar. Like light, they are both waves carrying energy through empty space at the amazing speed of 300 000 km/s and are both part of the **electromagnetic spectrum**. The only difference is the amount of energy that they carry. Of the three, light carries the most energy – but who is afraid of daylight?

a Name three parts of the electromagnetic spectrum.

b What do the waves carry through empty space?

Cooking with infrared

All warm and hot objects emit infrared radiation. Hot objects emit more radiation than warm objects and dull black objects emit more radiation than shiny white objects.

Most conventional ovens use infrared radiation to cook food. The designs of these ovens usually take into account the properties of infrared radiation:

- lots of infrared is emitted by hot surfaces
- more infrared is emitted and absorbed by black surfaces than by any other colour
- infrared is absorbed by the surface of an object so it gains heat energy
- infrared is reflected by shiny surfaces.

Of course, anything can be used to heat up the black surfaces in the oven. Gas, electricity, wood …

c **Why is black the best colour for the inside of an oven?**

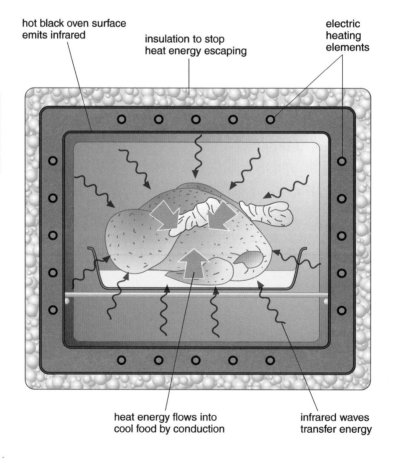

hot black oven surface emits infrared

insulation to stop heat energy escaping

electric heating elements

heat energy flows into cool food by conduction

infrared waves transfer energy

▲ *Infrared waves carry heat energy from the hot sides of the oven to the food*

Microwave meals

Have you ever wondered how a microwave oven works? When microwave radiation is absorbed by water the water heats up. Microwave ovens emit microwaves that penetrate about 1 cm into the food. They make the water molecules inside the food vibrate so much that they transfer heat energy to the food.

Your body contains a lot of water so you can get burned by microwave radiation if your body tissue absorbs it. Microwaves cannot go through metal but they can go through glass and plastic. This is why the glass window of a microwave oven has a metal grid to reflect the radiation back into the oven.

 Why is the inside of a microwave oven made of metal?

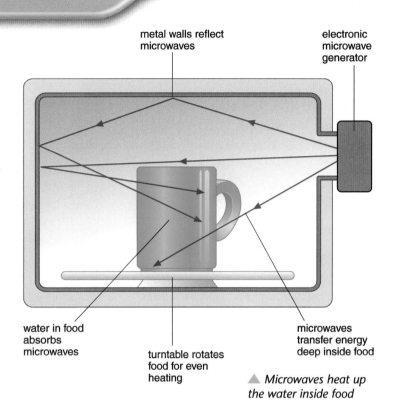

metal walls reflect microwaves

electronic microwave generator

water in food absorbs microwaves

turntable rotates food for even heating

microwaves transfer energy deep inside food

▲ *Microwaves heat up the water inside food*

Microwaves from mobiles

Mobile phones send out microwave signals. These are very low power – much less than used in microwave ovens. The information carried by the telephone microwaves is transmitted to mobile phone masts. To send a signal, your phone needs to be able to 'see' the mast. This is because microwaves cannot bend round corners or climb over mountains. Places that are not in line of sight with a mobile mast get poor signals. Mobile phone masts emit a lot more radiation than your phone does.

 Why might people worry about living close to a mobile phone mast?

some microwaves absorbed by the brain

mobile phone emits microwaves

◄ *Microwaves for long distance telecommunications are beamed out well above ground level so that people can't get in the way*

keywords

microwave •
electromagnetic spectrum

Are mobile phones fatal?

In1997, a team of scientists in Australia suggested that microwaves from mobile phones could cause cancer. The scientists bred 200 mice genetically disposed to getting cancer of the white blood cells. They then exposed half of them to pulsed digital microwaves close to the head and waited 18 months for cancers to develop.

They found that mice exposed to the microwaves had 2.4 times as many cancers as the control group. Should we be concerned?

The scientists said that their findings didn't mean that human beings ran the same risk as genetically adapted mice. More research was needed. They asked for funds to research the effect of microwaves on humans. Of course, you cannot do experiments of this sort directly with people.

So, the scientists compared the cancer rates for people who use mobile phones and those who do not. Up till now, no research has shown conclusively that using a mobile phone can give you cancer.

The chances of developing brain cancer over a whole lifetime are pretty small. So this sort of research is only going to deliver conclusive results if it involves a lot of people and monitors them over a long time. You can't rush this type of study.

The real problem is finding enough people who don't use mobile phones, but who are similar to people who do.

Questions

1 Suggest why brain cancer is the most likely form of cancer due to mobile phones.

2 The scientists had to make sure that the control group of people were matched for age, sex and lifestyle to the group who used mobiles. Explain why.

3 Suggest, with reasons, the best number of people in each group.

4 Describe how the scientists might run their experiment to find out if mobiles cause brain cancer.

5 A scientist proposes to repeat the mice experiment on cats. Another scientist says the experiment would be better carried out on monkeys. Which scientist do you think is right and why?

6 Why do you think the scientists did not suggest carrying out the same experiment on humans?

Everyday infrared

In this item you will find out

- the many ways that infrared is used
- about analogue and digital signals
- about optical fibres

Imagine what life would be like without television remote controls. If you wanted to change channel you would have to get up from your comfy seat and walk over to the TV.

Your TV remote uses infrared radiation to send signals between the remote and the TV. Unlike other radiation, such as microwaves, infrared cannot travel through solid walls, so it is ideal for this use.

So, how can you use two or more different remote controls in the same room?

Remote controls use a code to carry information. Each device has its own code and so it ignores infrared signals intended for other devices.

So the code used for controlling the TV is different to the code used for controlling a video recorder, DVD player or the automatic doors on a garage.

Infrared can also be used for sending data between computers or mobile phones that are a short distance apart.

a How does a remote control transfer information?

b The TV and DVD player have different remote controls. Why can you use the TV's remote without affecting the DVD player?

Infrared detection

People in buildings tend to be hotter than their surroundings. This means that they give off more infrared radiation than the objects around them. Some burglar alarms contain infrared sensors to detect the extra infrared emitted by people.

Some security lights also work in the same way. When they detect the infrared given off by a person, they switch on.

c Thieves sometimes try to fool burglar alarms by covering their faces. Why do you think they do this?

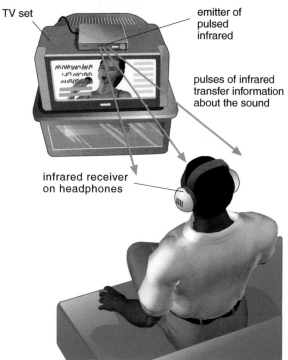

TV set

emitter of pulsed infrared

pulses of infrared transfer information about the sound

infrared receiver on headphones

Analogue and digital

Sound is an example of an **analogue** signal. The electrical signal from a microphone has a continuously variable voltage. The infrared communication sent by a TV remote works best with pulses – the beam of infrared is either on or off. This is an example of a **digital** signal.

Digital signals are usually written using 1 for on and 0 for off. So a typical digital signal might look like this on the printed page:

011000110010

If you want to convert analoque signals into digital signals before transmitting them, then you have to use a code.

d A motorcyclist can indicate a left turn with a raised arm or a flashing light. Which is the analogue signal?

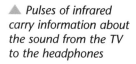
▲ Pulses of infrared carry information about the sound from the TV to the headphones

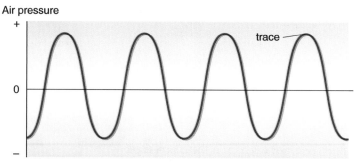
▶ Analogue signals can have any value

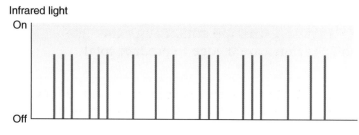
▶ Digital signals are on or off

Optical fibres

Infrared or visible light signals can pass through many kilometres of thin glass strands called **optical fibres**. Pulses of radiation fed in at one end of the optical fibre pass all the way to the other end. This allows optical fibres to transmit data quickly over large distances.

Even though the optical fibre is very transparent, the radiation can't leak out of the walls. Because the pulses are travelling almost parallel to the wall of the fibre, they are totally reflected back into the glass.

So, infrared or visible light has to follow the path of the optical fibre – even round corners! This is called **total internal reflection** (TIR).

 Why doesn't optical fibre need to be coated with a reflector?

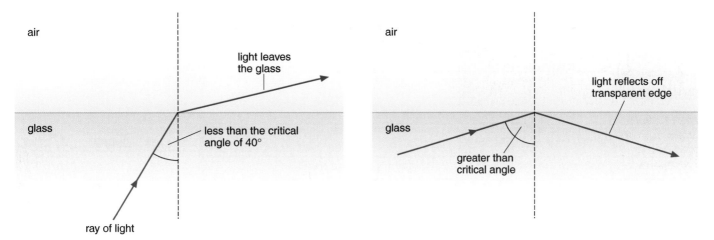

▲ Once light gets into the fibre it can't escape until it gets to the other end

Critical angle

A beam of light can only pass out of a transparent material (such as water, glass or perspex) if its angle of incidence is less than the critical angle. For glass, the **critical angle** is about 40°. Each material has its own value.

▲ Light can only escape if the angle of incidence is less than the critical angle

If the angle of incidence is equal to the critical angle, the light emerges from the material travelling parallel to its surface.

When the angle of incidence is greater than the critical angle then the light is internally reflected off the transparent edge to produce total internal reflection.

 The critical angle for water is 50°. Describe what happens to a beam of light in the water if hits the surface at an angle of 60°. What happens if the angle is only 40°?

keywords

analogue • critical angle • digital • optical fibre • total internal reflection

Pulses are perfect

The Internet relies on optical fibres to transfer the bulk of its information around the world. Without them, the communications network which allows every computer to be almost instantly linked to any other computer on the surface of the Earth would not exist.

In 1988 the first optical fibre was laid under the Atlantic Ocean. It could carry 8000 telephone conversations at once using pulses of infrared radiation. Infrared was selected over visible because its absorbtion by the fibre is minimal. However, after some hundreds of kilometres, the pulses became too weak to be clearly recognisable, so electronic amplifiers (called repeaters) were inserted at intervals to restore the pulses to their original intensity. Repeaters need to be very reliable because repairs are very difficult at the bottom of the ocean!

Since the information is carried in the timing of the pulses and not their intensity, none of it need be lost on its way across the sea. The use of digital coding allows for perfect, noise-free transmission of phone calls or anything else, such as pictures, video or text that can also be coded as a series of pulses.

Nowadays, optical fibre cable is not only under the sea but everywhere! Because the fibres are so thin, bundles of them have been laid in existing pipes, alongside water pipes, and electricity cables. They can't corrode, so they can even be laid in sewage pipes!

Questions

1 Why do long-distance optical fibres use infrared?

2 What is a repeater? Why is it necessary?

3 Why are long-distance phone calls through optical fibre noise-free?

4 Describe three different types of information that can be sent down optical fibre.

5 The optical fibre network is growing rapidly. What properties of the fibre make this possible?

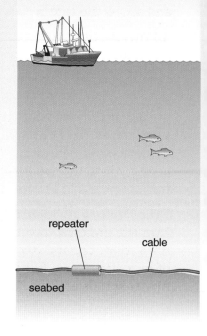

▲ *Repeaters along an undersea cable*

Cut the cable

In this item you will find out

- about wireless technology and its advantages

- about reflection and refraction

- why digital radio has advantages over analogue radio

Your mobile phone means that you need never be out of touch. Wherever you are, whatever you are doing, people can contact you. You can exchange information in different ways – text, sound or pictures. Only when the battery (or your credit) runs out do you realise the freedom that a mobile phone gives you!

The mobile phone is only the start of the wireless revolution.

Cables of wire for ordinary computers are expensive to install, and the need to be plugged in keeps you in one place all the time. But wireless laptop computers are becoming widely available, with new wireless networks appearing all the time. This means you can take your laptop anywhere and access the Internet or your emails 24 hours a day.

All this is possible because wireless technology uses electromagnetic radiation to send and receive data.

But what about the security of wireless networks? The waves spread out in all directions, so anyone can listen in! However, the power of these transmissions is usually kept low, so they can't travel far before becoming too weak to be picked up. If signals from different networks do stray into the same area, then careful selection of the wavelengths allocated to each network usually avoids confusion for the receiver.

Coming alongside the wireless revolution is the digital revolution. The use of digital coding allows communication of more types of information (sound, text, pictures) at higher speeds and quality than ever before.

Radio reception

Radios also use wireless technology. Radio stations broadcast their signals from transmitter masts to our homes. Locating the masts in high places increases the range of the radio waves. You can only get good reception if the mast is above your horizon. If you are too far away from the mast, then the signal at your radio will be too weak for you to hear the sound clearly.

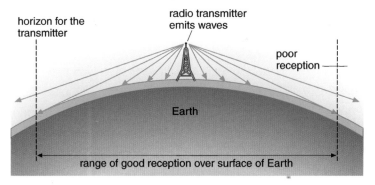

horizon for the transmitter

radio transmitter emits waves

poor reception

Earth

range of good reception over surface of Earth

▲ A high location for the transmitter increases its range

The transmitter emits lots of different radio waves at once, each with a different frequency. Each radio station is allocated its own set of **transmission frequencies** (called a channel) for coding its sound. When you change station on your receiver, you are changing the range of transmission frequencies that your radio is receiving.

a What would happen if two different channels tried to use the same set of transmission frequencies?

b Broadcast radio, like BBC Radio 1, is not the only use for radio waves. Suggest what else they can be used for.

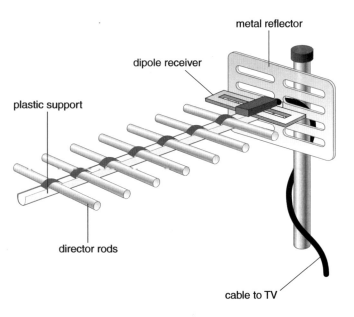

metal reflector

dipole receiver

plastic support

director rods

cable to TV

Reflection

Radio waves are electromagnetic waves. So radio waves are reflected by metal objects. Televisions receive radio waves. By putting a reflector behind your TV aerial, you can improve the reception. Of course, for this to work, the aerial must point straight at the transmitter.

Satellite TV aerials use curved reflectors to increase the strength of the signal from the satellite in space.

c Transmitters which send radio signals to satellites in space have large round dishes behind the aerial. Why?

▲ TV reception can be improved by reflectors behind the aerial

Refraction

Remote controls use light-emitting diodes (LEDs) to produce their pulses of infrared. As the infrared leaves the plastic and enters the air, it changes direction. This is called **refraction.** All electromagnetic waves can be refracted. The plastic of the LED is shaped like a lens, forcing most of the infrared to go in one direction.

Refraction can also be used at the receiver. The correct shape of plastic focuses the pulses of infrared onto the detector surface, increasing the strength of the signal.

 Why do you think LEDs emit infrared over a small range of angles instead of just a single direction?

Waves refract because they change speed when they enter or leave a particular material. The speed of a wave depends on what it is passing through.

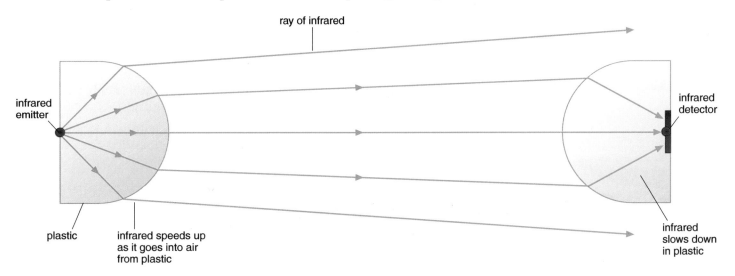

ray of infrared

infrared emitter

plastic

infrared speeds up as it goes into air from plastic

infrared detector

infrared slows down in plastic

▲ *Refraction happens when the wave speeds up or slows down*

As radio waves pass through the Earth's atmosphere they are refracted as they move from high density air into low density air. This allows radio waves to travel over the horizon.

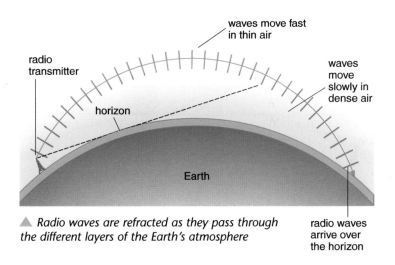

waves move fast in thin air

radio transmitter

horizon

waves move slowly in dense air

Earth

radio waves arrive over the horizon

▲ *Radio waves are refracted as they pass through the different layers of the Earth's atmosphere*

Examiner's tip

Refraction and diffraction are easily confused.

keywords

refraction • transmission frequencies

Radio Rapster goes digital

Winston runs Radio Rapster, a local radio station broadcasting at about 103 MHz.

'I can only reach an audience within about fifty miles of the transmitter.' says Winston. 'The radio waves won't bend over the horizon. It's a shame, because we have a good following and we could go national. But all of the frequencies are already used in the surrounding areas, so there's no chance of expansion.'

Winston broadcasts in the Frequency Modulation (FM) band. This uses radio waves of frequencies between 88 and 104 MHz. Each station in this band broadcasts stereo signals over the whole audible range, using analogue coding.

But, reception isn't always perfect, and only about a dozen different stations can broadcast in each area. Each has a separate set of frequencies, so that there is no confusion for the receiver.

Winston decides to apply for a licence to broadcast in the new Digital Audio Broadcasting (DAB) band. It uses digital coding, switching on and off radio signals in the band between 218 MHz and 230 MHz. This improves the reception, with less background noise.

The sound information is spread over 1536 different channels, allowing each station to broadcast over the whole country. Almost a hundred different stations can broadcast in the same area!

Questions

1 What does FM mean? What are the disadvantages of FM radio?

2 What would happen if two radio stations near to each other were allowed to broadcast on the same set of FM frequencies?

3 Why can Radio Rapster in the FM band only broadcast to a small area of the country?

4 What does DAB mean?

5 What are the advantages of DAB compared with FM broadcasting?

Making light work

In this item you will find out

- the features of a transverse wave

- how pulses of light can be used to send messages in code

- what lasers can do

How fast can information travel? At the speed of light!

As you read these words, information is being transferred from the page to your brain by light. Light from the room is reflected off the paper but absorbed by the ink. The reflected light is detected by your eyes and builds up an image of the words in your brain. Your brain decodes the letters and reads the message. The light carries information from the page to your eyes in almost no time at all. It takes your brain a lot longer to work out what the message is.

For most of recorded history, a written or spoken message could only travel as fast as the human or animal that was carrying it.

In 1793, Claude Chappe used an optical telegraph to send a 150 word message over a distance of about 30 kilometres in only 11 minutes. The light which carried information about the positions of the telegraph arms only took 0.0001 seconds to get from transmitter to receiver.

You can now get a message to anywhere on Earth in, at most, a few seconds. Since all electromagnetic waves move at the same speed, it doesn't matter which one you use.

▲ *A Chappe-style optical telegraph*

a Describe how the telegraph station in the picture sends a message. How far can it travel?

Amazing fact

Light takes eight minutes to get from the Sun to Earth.

Wave properties

Electromagnetic waves are transverse **waves** which vibrate up and down.

The diagram shows what a transverse wave looks like. It is made up of a series of **crests** and **troughs**.

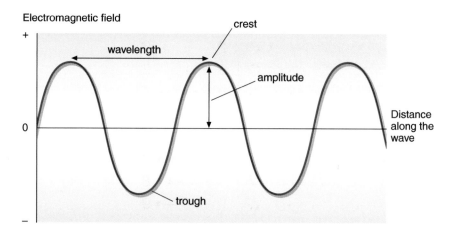

Here are the features of a transverse wave:

- the **frequency** is the number of waves produced by the source each second
- the **wavelength** is the distance from one wave crest to the next
- the **amplitude** is the height of the crests.

The frequency of a light wave is measured in hertz (Hz). The wavelength is measured in metres (m).

You can calculate the speed of a wave by using this equation:

speed = frequency × wavelength

All electromagnetic waves travelling in empty space have the same speed. This means that if you increase the frequency of a wave, the wavelength will automatically decrease.

 A wave has a wavelength of 20 cm and a frequency of 11 Hz. Calculate its speed.

Lasers produce a special type of light. The beam is narrow and very intense. The waves all move in the same direction and all have the same frequency.

Communicating with light

Light has been used for centuries as a way of sending messages from one place to another. Using light enabled people to communicate with each more quickly but it meant that they had to send messages in code.

In 1832, Samuel Morse invented his famous code of dots and dashes called **Morse code**.

It was first used to send messages through wires by pulsing the current in them on and off. Sometimes, the current controlled a pen on the end of an electromagnet, leaving a trail of dots and dashes on a strip of paper moved past the pen.

A	•–	A	N	–•	N	0 —————	0
B	–•••	B	O	———	O	1 •————	1
C	–•–•	C	P	•——•	P	2 ••———	2
D	–••	D	Q	——•—	Q	3 •••——	3
E	•	E	R	•–•	R	4 ••••—	4
F	••–•	F	S	•••	S	5 •••••	5
G	——•	G	T	–	T	6 –••••	6
H	••••	H	U	••–	U	7 ——•••	7
I	••	I	V	•••–	V	8 ———••	8
J	•———	J	W	•——	W	9 ————•	9
K	–•–	K	X	–••–	X		
L	•—••	L	Y	–•——	Y		
M	——	M	Z	——••	Z		

◀ Morse code

Pulses of light can also use Morse code, or any other sort of code, to transfer information. Lasers are particularly useful for this. They can:

• read bar codes on products in supermarkets
• extract information about sound stored in CDs
• link computers with optical fibres.

 Suggest one advantage of using light instead of wires to send messages in Morse code.

◀ A laser is used to scan these bar codes

Levelling with light

Steve works on a building site. He has 48 hours to get the site level and flat, ready for laying the foundations. The derelict factories have been torn down, leaving heaps of rubble and holes all over the place. He walks over to a carefully cleared circle in the centre of the site and sets up a tripod. At the top of the tripod, above eye-height, he sets up a laser on a special rotating platform. Having checked the platform is perfectly level, Steve activates the motor under the platform, and the laser starts spinning round. It spins round exactly 13 times a second.

Steve speaks into his walkie-talkie, and six bulldozers cough and splutter to life. Each has a vertical rod next to the cab, with light sensors along its length. This tells each driver the height of the ground under his bulldozer, compared with the spinning laser. So as they move around the site, they know whether to scrape material off the surface or add more. To start with, they push rubble into holes. Then they go to the edge of the site, collect some gravel and use it to fill in dips.

As the air fills with dust, the spinning red laser beam becomes visible as it scatters off the particles suspended in the air. It looks like a thin red sheet about two meters above the ground – very impressive at night. By next morning, the task is complete and Steve can turn off the laser. The site is now flat and level …

▲ The spinning laser is a useful tool in levelling the land

Questions

1 Suggest why Steve uses laser light.

2 Why do you think the laser is set above eye-height.

3 Suggest why the light sensors on the bulldozers will only respond to light which pulses at 13 Hz.

4 The bulldozer cabs have dark glass in the windows. Suggest why.

5 Lasers are also used to make sure that road tunnels through mountains are straight. Suggest how this works.

Killer waves

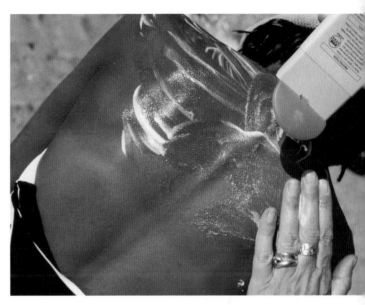

In this item you will find out

- about the dangers of exposure to ultraviolet radiation

- why trapped infrared radiation is causing climate change

- how the energy of earthquakes is destructively spread through waves

When you go to the beach, you need to make a decision. Do you keep out of the Sun and avoid the risk of skin cancer?

Or do you expose your skin so that it eventually gets tanned and not worry about the consequences?

The beach used to be the place to get a nice tan. Now that chemical pollution has thinned the ozone layer, too much **ultraviolet** from the Sun is getting to ground level. So you ought to slap on the sunblock and stick on a hat!

The beach is threatened in other ways. It could soon be flooded by rising sea levels caused by **global warming**.

Pollution from burning fossil fuels is preventing infrared radiation from leaving the Earth, making its surface warmer. The ice caps are melting and the sea is expanding ...

Some threats have always been there. The destructive waves on land and sea from **earthquakes** are not new. They can strike at any time ...

The tsunami in 2004 was caused by an earthquake at sea ▶

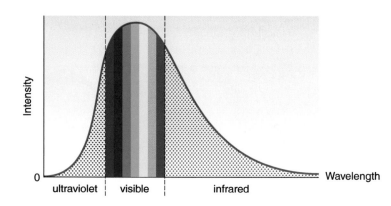

Intensity

ultraviolet | visible | infrared

Wavelength

0

▲ *The Sun emits most radiation from the visible part of the electromagnetic spectrum*

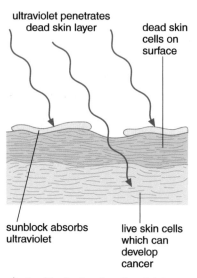

ultraviolet penetrates dead skin layer

dead skin cells on surface

sunblock absorbs ultraviolet

live skin cells which can develop cancer

▲ *Sunblock absorbs ultraviolet before it reaches live cells*

Solar radiation

The surface of the Sun emits three types of electromagnetic radiation:

- long wavelength infrared radiation which warms up the planet
- medium wavelength visible light which is vital for photosynthesis in plants
- short wavelength ultraviolet radiation which is bad news for living organisms.

Ultraviolet kills

The short wavelength of ultraviolet (UV) radiation means that it has enough energy to smash chemical bonds. It can easily damage molecules (such as DNA) in living cells. So live skin cells exposed to ultraviolet can develop skin cancer. Many people who live in sunny climates have dark skins. Having a dark skin reduces the risk of getting cancer as dark skin absorbs more ultraviolet than pale skin. This means that less ultraviolet light reaches the live skin cells below the dead surface.

When people with pale skins are exposed to ultraviolet radiation from the Sun they often develop tans. However, sunburn is a painful sign that ultraviolet is damaging you! You should avoid this by putting sunblock on exposed skin. The **sun protection factor** (SPF) of a sunblock indicates how much UV it absorbs. For example, an SPF of 8 allows you to stay up to eight times longer in the Sun without burning. The higher the SPF, the more the risks of damage are reduced.

a Why are people with dark skin less likely to develop skin cancer than people with pale skin?

b Jane gets sunburn after only half an hour on the beach with no sunblock. How long would it take if she used sunblock with an SPF value of 6?

Climate change

Two things keep our Earth at a comfortable temperature:

- the Earth absorbs light and infrared from the Sun, heating it up
- the Earth radiates infrared into Space, cooling it down.

Humans and nature can both alter this delicate balance, causing the temperature of the planet to gradually change:

- dust from volcanoes reflects radiation from the Sun, which causes the Earth to cool down
- dust from factories reflects radiation from cities, which causes warming of the Earth
- using more fossil fuels means creating more CO_2, which stops infrared radiation escaping from the Earth, which results in warming
- burning forests also means creating more CO_2, which causes the Earth to warm.

A warmer planet means that sea levels will rise and some plants and animals will become extinct as the climate changes.

energy from Sun as light

empty Space above atmosphere

gases in the atmosphere are transparent for light but not for infrared

infrared emitted by hot ground

infrared trapped by carbon dioxide in atmosphere

ground transfers light to heat energy

Earth

c **What are humans doing to warm up the Earth? What effect will it have?**

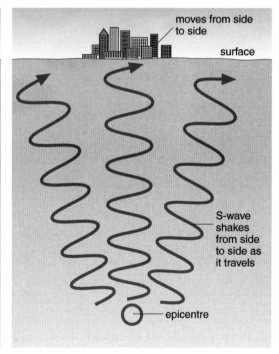

▲ *Damage from a recent earthquake*

Violent Earth

Earthquakes release a lot of energy very suddenly, deep underground. The energy is carried away by seismic shock waves which can travel inside the Earth. One type of wave, called a **P-wave**, pushes the ground back and forth as it moves along. It is a **longitudinal** wave.

The other type of wave shakes the ground from side to side. It is called an **S-wave** and it is a transverse wave.

When they reach the surface, both of these shock waves can cause a lot of damage.

S-waves can travel through solids but not through liquids so they are stopped by the liquid core of the Earth. P-waves can travel through solids and liquids.

Because P-waves can travel across the liquid core, they can be detected by **seismometers** on the other side of the Earth. P-waves also travel faster than S-waves.

d **What is the difference between an S-wave and P-wave?**

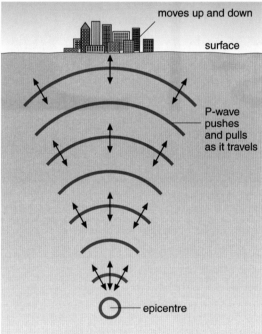

moves up and down

surface

P-wave pushes and pulls as it travels

epicentre

moves from side to side

surface

S-wave shakes from side to side as it travels

epicentre

▲ *S-waves and P-waves have different effects on the ground that they pass through*

▲ *The seismometer shows a trace produced by an earthquake*

keywords

longitudinal • P-wave • S-wave • earthquake • ultraviolet • sun protection factor • seismometer

Choosing sunblock

Siân and Michelle are going on holiday to Tunisia. They plan on spending a lot of time on the beach enjoying the Sun. Today they are going shopping for sunblock. There are a lot of different types to choose from.

Michelle picks up the cheapest sunblock. 'This one will do.' she says. 'I'm not spending lots of money on it.'

'But that one only has a low SPF.' says Siân.

Michelle isn't sure. 'What does that mean?'

'The higher the SPF the longer you can stay in the Sun without getting burned. A high SPF reduces your risk of getting skin cancer. An SPF of 15 means you can stay 15 times longer in the Sun without getting burned than if you weren't wearing any sunblock.'

Questions

1 What are the advantages of using sunblock with a high SPF?

2 How much longer can Michelle stay in the Sun if she is wearing sunblock of SPF 25 than if she is wearing none at all?

3 The sunblock that Michelle first chose had an SPF of 8. People can usually stay in the hot Sun for 15 minutes without burning. How long could Michelle stay in the Sun wearing this sunblock?

P1a

1 Complete the sentences. Choose from:

J °C W

The heat energy of an object is measured in ____(1).

Its temperature is measured in ____(2). [2]

2 Heat energy is added to a lump of ice at −10 °C until it becomes steam at 200 °C. Put the sentences in the correct order.

A the liquid gets hotter   B the solid gets hotter
C the gas gets hotter   D the solid melts
E the liquid boils [4]

3 Complete the sentence.

The specific heat capacity of a material is the ____(1) needed to raise the temperature of one ____(2) by one ____(3). [3]

4 Explain what the specific latent heat of material is. [2]

5 A hot object is placed in contact with a cold one.

   a Which way will energy flow between them? [1]
   b What will that flow of energy do to the temperature of the objects? [3]

6 Complete the sentences. Choose from:

heat   rises   temperature   latent
liquid   32 °C   37 °C

Chocolate melts at 32 °C. Your mouth has a ____(1) of 37 °C. When solid chocolate is put in your mouth, its temperature ____(2) as it absorbs ____(3) energy from you.

When the chocolate reaches its melting point of ____(4), its ____(5) stops rising as it absorbs ____(6) heat and turns from a solid to a ____(7). The liquid finally reaches a temperature of ____(8) as you swallow it. [8]

P1b

1 Here are some materials. Which ones are good conductors?

copper   iron   plastic   wood [2]

2 Describe three ways of reducing heat energy loss from a building [3]

3 Insulating the roof of a house costs £1000, saving £200 a year in heating bills. Calculate the payback time for insulating the roof. [2]

4 Some meat is being cooked in a microwave oven. For every 60 J of electricity transferred into the oven, only 45 J of heat energy transfers into the meat. Use the rule

$$\text{efficiency} = \frac{\text{useful energy output}}{\text{total energy input}}$$

to calculate the efficiency of the oven. [2]

5 Complete the sentences. Choose from:

conduction   convection   radiation

Heat energy is lost through the solid walls of a house by ____(1). As warmed air rises from the roof, heat energy is lost through ____(2). Increasing the temperature inside the house increases the amount of heat lost by ____(3) through the windows. [3]

6 Explain the following statements about heating houses. Use the words **conduction**, **convection** and **radiation** in your answers.

   a Keeping the windows open in winter makes a house colder. [2]
   b Lowering the thermostat setting in winter saves money. [2]
   c Painting a house white keeps it cooler in summer. [2]
   d Shiny foil placed on the wall behind radiators reduces heating bills. [2]
   e Emptying the loft makes the house colder. [2]

P1c

1 Complete the sentences. Choose from:

absorbs   falls   floats   reflects   rises

An electric heater heats the air in the room. The hot air ____(1) and is replaced by the cold air which ____(2). Shiny foil on the wall behind the heater ____(3) heat back into the room. [3]

2 Match the start and end of the sentences.

1 a cavity wall   a reduces heat loss by conduction
2 drawing curtains   b reduces heat
    across windows   loss by radiation
3 planting trees   c reduces heat loss
    around a house   by convection [3]

3 Insulating materials are usually full of holes. Explain why these holes make the materials poor conductors of heat. [2]

4 Heat energy can be saved by painting a house white. State and explain the type of heat transfer that this reduces. [2]

5 A pot of ice cream is wrapped in plastic bubble wrap before being surrounded by aluminium foil. Explain how this helps to keep the ice cream cold in a warm room.

Your answer should include:

- the direction of flow of heat energy
- the effect of the foil
- the effect of the bubble wrap.   [3]

P1d

1 Which one of these waves is not part of the electromagnetic spectrum?   [1]

infrared   **microwaves**   **sound**

2 Complete the sentences. Choose from:

black   **infrared**   **microwave**   **red**   **shiny**

A gas oven heats food with ____(1) radiation.

The best colour for the inside of the oven is ____(2).

Food that is heated with ____(3) radiation must contain water to absorb the radiation.   [3]

3 Complete the sentences.

Microwaves are reflected by ____(1) but transmitted by ____(2).

In an oven, microwaves are absorbed by ____(3) in the food.   [3]

4 Microwaves carry telephone messages from one place to another. Describe and explain the limits on how far apart the two places can be.   [3]

5 Compare the different ways in which food is heated in a microwave oven and a gas oven.   [3]

6 It has been suggested that microwaves from mobile phones don't give people cancer of the brain. Describe how scientists could collect evidence to support this suggestion.   [4]

P1e

1 Describe two uses of infrared radiation.   [2]

2 There are two types of signal used in communication. One is analogue. What is the other?   [1]

3 Infrared sensors are used to detect burglars. What aspect of the burglar's body do they detect? Choose from:

colour   **heat**   **smell**   **sound**   [1]

4 Optical fibres allow the rapid transmission of data using pulses of light.

- **a** Describe how light passes from one end of a fibre to the other.   [2]
- **b** What type of signal are the pulses of light?   [1]
- **c** The data being transmitted is speech. What type of signal is speech?   [1]

5 Light in glass hits the surface below the critical angle. Which of the sentences best describes what happens to the light.

A all of it passes out of the glass into the air
B some passes into the air, the rest is reflected back into the glass
C all of it is reflected back into the glass   [1]

6 Complete the sentences. Choose from:

absorbed   **critical**   **distance**   **glass**
reflects   **refracts**   **transparent**   **escape**

Optical fibres are made of ____(1) which is ____(2) to infrared. This allows the infrared to travel a long ____(3) before it is completely ____(4). The infrared hits the edge of the fibre at more than the ____(5) angle, so it all ____(6) and none ____(7) out of the glass.   [7]

P1f

1 Complete the sentence. Choose from:

colour   **electromagnetic**   **fibres**   **sound**
space   **wires**

Wireless communication uses ____(1) radiation travelling in ____(2).   [2]

2 State two advantages of wireless communications for someone using a computer.   [2]

3 Name three common uses of wireless technology.   [3]

4 Radio waves travelling away from the Earth can sometimes follow a curved path back to Earth. What is the name of this effect?   [1]

5 Why are different radio stations in an area not allowed to broadcast with the same frequency?   [1]

6 What is the difference between refraction and reflection? Give an example of each as part of your answer.   [4]

P1g

1. Draw two cycles of a wave. Label a trough and crest. Indicate the distances which represent the amplitude and the wavelength. [4]

2. Which of these produces a narrow intense beam of light?

   A light emitting diode   B light bulb
   C match   D laser [1]

3. Complete the sentence. Choose from:

   code   mirror   colour   shade

   When light is used to send a message, a ____(1) must be used. [1]

4. A sound wave with a frequency of 680 Hz has a wavelength of 0.5 m. Use the rule

   speed = wavelength × frequency

   to calculate the speed of the wave. [2]

5. Match the start and end of the sentences

   1 the frequency   a the height of a crest
   of a wave is
   2 the amplitude   b the distance between troughs
   of a wave is
   3 the wavelength   c the number of waves in a
   of a wave is   second [2]

6. Describe how light can be used to send messages in Morse code. [2]

7. Ships can use strings of flags to send messages to each other.

   They can also use beams of light.

   a Explain how a message can be sent along a beam of light. [2]
   b Suggest why strings of flags are not used when the ships are part of a naval battle. [1]
   c Ships in battle use radio waves for communication instead of light. Suggest why. [1]

8. Describe **two** uses of lasers. Your answer should include a description of the properties of laser light which allow it to be used in this way. [4]

P1h

1. Which of these instruments can be used to detect an earthquake?

   A joulemeter
   B seismometer
   C thermometer [1]

2. Complete the sentences. Choose from:

   **alcohol   cancer   infrared   nuclear   spots
   sunblock   ultraviolet   warts   water**

   Sunbathing can give you ____(1) due to the ____(2) radiation.

   You can prevent this by putting ____(3) on your skin.[3]

3. Complete the sentences.

   There are two types of seismic wave, called ____(1) and ____(2).

   They are produced by earthquakes. ____(3) are transverse and can only travel through ____(4).

   ____(5) are longitudinal, so they can travel through both solids and____(6). [6]

4. Explain why people with darker skins have less risk of skin cancer. [3]

5. A sunblock has an SPF of 4. What does this mean? [2]

6. Explain how the action of volcanoes can result in global cooling. [2]

7. State the causes of global warming. [3]

8. Complete the sentences. Choose from.

   **carbon dioxide   ground   heat   infrared
   more   Sun   temperature**

   Light energy from the ____(1) passes straight through our atmosphere.

   It is absorbed by the ____(2) and transerred to ____(3) energy.

   The warmed ground emits ____(4) which is partly reflected by ____(5) in the atmosphere.

   Increasing the amount of carbon dioxide results in ____(6) infrared from the ground being trapped, raising the ____(7) of the atmosphere.

   [7]

P2 Living for the future

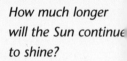

Where does my electricity come from? Can I choose how it is made?

- The supply of the fuels which are used to make most of the electricity you use today will soon run out. There is only so much oil, coal and gas in the ground, and they are not being replaced. Nuclear power may provide one answer to this problem and the Sun another.

Oil is going to run out soon. What is going to replace it?

How much longer will the Sun continue to shine?

- We may need to use resources on other planets to supplement the ones we are using up on Earth, but space travel is dangerous. At least we now know that the Sun will carry on providing heat and energy for another five billion years!

What you need to know

- Moving objects have kinetic energy.

- Plants transfer light energy into chemical energy.

- Non-renewable fuels will not be replaced once they have all been used up.

- Magnets are surrounded by energy fields which can interact with other magnets.

Sunlight and wind power

In this item you will find out

- how photocells make electricity from sunlight

- how solar panels and wind turbines can use the Sun's energy

Do you ever think about where the energy you use comes from? Your use of energy is helping to change our planet – possibly for the worse!

Most of the electricity you use is made by burning gas. This adds carbon dioxide to the atmosphere, and leads to heat from the Sun being trapped in the Earth's atmosphere.

So, each time you plug your mobile phone into the mains to recharge it, you are heating up the Earth. Many people are convinced that this increase in carbon dioxide is bad for the planet.

Here are three reasons why:

- it makes the weather more violent
- the sea expands flooding the land
- the climate changes, which may lead to some plants and animals becoming extinct.

Some governments have agreed that this must stop. The race is on to find ways of making electricity which don't damage the environment, such as using renewable energy from the Sun.

The Sun supplies energy to the Earth in the form of heat and light. It is a stable source of energy because the energy is always available.

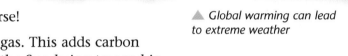

▲ Global warming can lead to extreme weather

▲ This power station uses gas as its fuel

Photocells

A **photocell** works by transferring light energy into electrical energy. It produces **direct current** (DC), a current that flows in the same direction all the time. A photocell does not need power lines or a source of fuel so it is ideal for using in remote places. The amount of electricity that a photocell produces depends on the size of its surface exposed to sunlight. Each square metre of photocell can give an electrical power of up to 50 W. This is enough to run a laptop computer. The only problem with photocells is that they do not produce electricity when there is no light – either at night or in bad weather.

Emma's calculator is powered by a photocell. The calculator doesn't need much power, so it can have a small photocell.

Winston needs a much larger photocell to power his laptop computer. The photocell charges up a battery so that his laptop still works when it gets dark.

▲ Emma's calculator runs off light

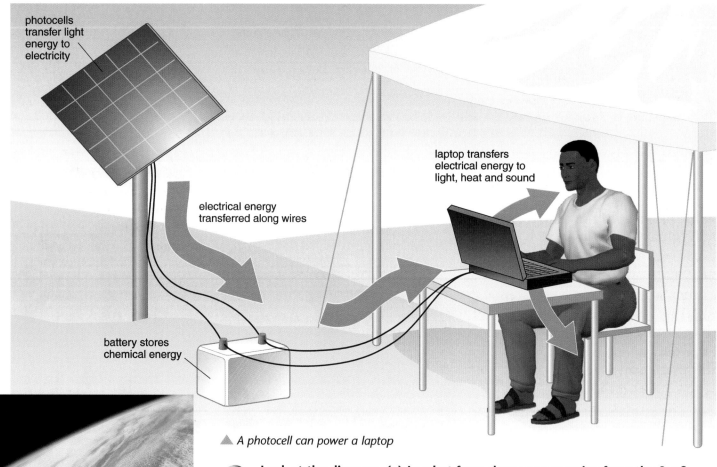

photocells transfer light energy to electricity

electrical energy transferred along wires

laptop transfers electrical energy to light, heat and sound

battery stores chemical energy

▲ A photocell can power a laptop

a Look at the diagram. (a) In what form does energy arrive from the Sun?

(b) What form of energy comes out of the photocells?

(c) In what form is it stored in the battery?

Photocells are good because they last a long time and are very tough. If you keep the photocell surface clean then no other maintenance is needed. They use a renewable energy source and do not cause pollution.

▲ This space station uses photocells

b Suggest two reasons why many satellites are powered by photocells.

Solar cooking

The Sun's light can be used to heat up an object. The light is absorbed by the surface of the object and transferred into heat energy. This light is a renewable source of energy because it does not run out.

Many people in the world burn wood to cook their food. Using the energy from the Sun instead makes good sense. But you need to use a curved mirror to focus the light onto the pot holding the food. The pot absorbs the light energy and heats up, cooking the food inside it.

c **Suggest two reasons for using a solar cooker. Give one disadvantage.**

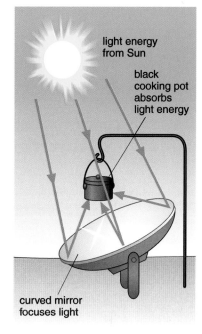

▲ *The mirror focuses the light onto the cooking pot*

Heating houses

Rashid has a solar panel on his roof. It has water pipes under a layer of glass. The light energy from the Sun heats up the water in the pipes to provide him with hot water.

He also has a wall of glass on the sunny side of his house. This heats up the air in his house. Rashid pays nothing for sunlight!

d **Why do you think the pipes in a solar panel are painted black?**

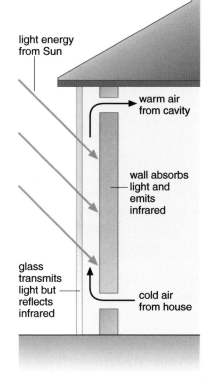

▲ *Infrared is trapped between the glass and the wall, heating up the air*

Wind power

Heat energy from the Sun sets air in the atmosphere moving as convection currents (wind). A **wind turbine** transfers the **kinetic energy** of moving air to electrical energy. Large turbines catch more air than small ones, so they generate more electricity. A really big turbine can supply enough electricity for a thousand homes – but only when the wind is strong enough!

e **Why does a large wind turbine make more electricity than a small one?**

▲ *A large wind turbine*

Amazing fact

The UK has a target of making 10% of its electricity from renewable sources, such as wind turbines, by the year 2010.

keywords

direct current • kinetic energy • photocell • wind turbine

Siting the sign

'It can't go there' says Pete 'The tree's in the way. It won't get enough light if we put it there.'

'Well, I can't move the road', says Sunita. 'This sign needs to go exactly there so that motorists see it before they get to the bend.

You'll just have to get rid of the tree. The chainsaw's in the van.'

Pete and Sunita are installing a road sign on a remote stretch of road, well away from towns or villages. The sign lights up whenever it detects an approaching car, to warn the driver of the sharp bend ahead. This means that it needs an electricity supply.

It gets its power from an array of photocells and a wind turbine. These charge a battery built into the sign, so that it always has power.

This is far cheaper than laying 10 km of cable alongside the road, all the way from the nearest village.

'Have you got the compass?' asks Sunita, 'We need to get the photocell pointing due south.'

Once the photocell is correctly aligned, Pete goes up the ladder for one last check of its angle to the horizontal.

'If it is too flat, dust and muck won't get washed off by the rain. If it is too steep, it won't catch enough sunlight when the Sun is high in the sky. The angle needs to be just right.'

Questions

1 Suggest why Pete needed to cut down the tree.

2 Why has the road sign got two sources of power?

3 Why is the battery necessary?

4 Describe and explain the need to get the photocell pointing in the right direction.

5 Explain why Pete and Sunita's road sign isn't run off mains electricity, even though it is much more reliable than turbines or photocells.

Making electricity

In this item you will find out

- about the dynamo effect

- how electricity is produced and distributed

- how transformers are used

Imagine a life before electricity became widely available. No computers. No microwave ovens. No refrigerators. No radio. No vacuum cleaners. The list is almost endless.

Most of us are surrounded by equipment which runs off electricity. Here are some of the things it does for us:

- preserves and cooks our food
- heats our houses
- cleans our floors
- entertains us with sounds and pictures
- carries our messages to other people.

Electricity is everywhere. At the flick of a switch, a host of different useful devices come to life. But where does it come from? What environmental price are we paying for it?

Most of your electricity is made in large power stations. These are usually sited well away from large centres of population, often within easy reach of their fuel supply. All of them have an effect on the environment in some way or other. Some produce radioactive waste. Others add carbon dioxide to the atmosphere. All of them can pollute the surrounding area with their waste heat. But does it have to be like this?

a List ten different items in your home which use electricity.

b How do power stations affect the environment?

Magnetism makes electricity

You can generate electricity by either moving a magnet past a coil of wire or moving a coil of wire past a magnet. This is called the **dynamo effect** and produces a voltage in the coil of wire.

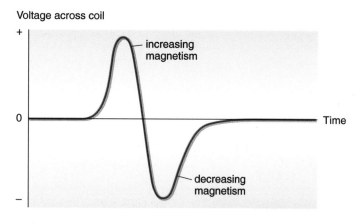

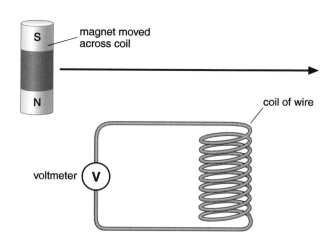

▲ When the magnet is moved past the coil a voltage is produced

The size of the voltage can be increased by:

• having more coils of wire
• wrapping the coil around a magnetic material, such as iron
• moving the magnet or the coil faster
• using a stronger magnet.

Generators

A simple **generator** uses a spinning magnet to transfer kinetic energy into electrical energy. It is made from a coil of wire wound round an iron core. As the magnet spins, it alters the magnetic field in the iron core. Each time the magnetic field changes, a voltage appears across the coil of wire. Iron is used for the core because magnetic fields travel easily through it.

 Suggest three ways of increasing the voltage across the generator coil.

Large generators use electromagnets. These are made with coils of wire wrapped around cylinders of iron. The electricity for the electromagnets comes from small generators with spinning permanent magnets made from steel.

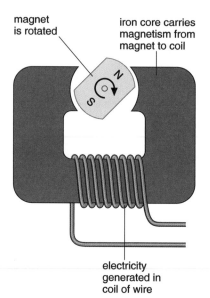

▲ How a simple generator works

AC and DC

The voltage from a generator is always changing. It rises as the magnetism increases, but then falls again as the magnetism decreases. The direction of the induced voltage changes, so the direction the current flows also changes from one direction to the opposite direction. In this way, generators produce **alternating current** (AC). The voltage from a battery, which is illustrated on the next page, is different. It has a steady value, producing direct current (DC).

 Describe the difference between AC and DC.

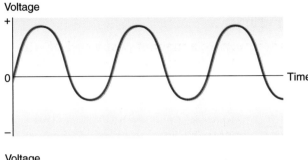

Voltage / + / 0 / − / Time

Voltage / + / 0 / − / Time

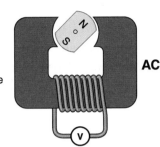

AC

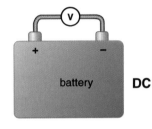

battery **DC**

◀ *Voltage-time graphs for a generator and a battery*

▲ *Inside a power station*

Spinning with steam

Power stations extract energy from their fuel as heat energy. High pressure steam from boiling water spins turbines, allowing the generator to produce electricity.

 How does energy in fuel become electricity?

Wasting heat

Not all of the energy in the fuel becomes electricity. Quite a lot of it is wasted as heat energy. Coal-fired power stations typically waste about 60% of the chemical energy in coal. Gas-fired power stations waste less energy.

 Power stations are often built by rivers. Suggest two reasons why.

100 J of chemical energy in coal → power station → 40 J of useful electrical energy

60 J of waste energy

▲ *Energy flow in a coal-fired power station*

Electricity across the UK

All the power stations in the UK feed their electricity into the **national grid**. This is a network of cables which span the whole country, carrying electricity to the consumers in their homes, their places of work or anywhere else they need it. The grid carries electricity at a very high voltage to reduce the heating up of the cables and keep costs down. So before a power station can feed its electricity into the grid, it has to be passed through a **transformer** to increase its voltage to several hundred thousand volts. Another transformer between the grid and the consumer reduces the voltage to a safer 230 V.

 Why do we use transformers to connect the power stations to the national grid and to consumers?

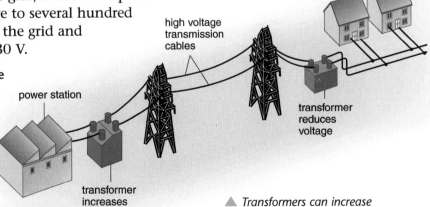

high voltage transmission cables

power station

transformer increases voltage

transformer reduces voltage

▲ *Transformers can increase or decrease voltage*

Green power

The national grid allows each consumer to buy their electricity from any supplier. All the grid does is to carry the electricity from where it is made to where it used. So, a business based in London could choose to buy its electricity from a nuclear power station on the north coast of Scotland. Or it could choose to buy electricity from nuclear power stations in France, closer to home.

▲ *Some electricity in the UK comes from French power stations*

The national grid gives each consumer a choice. Do I buy the cheapest electricity that I can find, or do I buy electricity which is less harmful for the environment – green electricity? Many of us live in towns or cities where it would be difficult to generate our own electricity from sunlight or wind power. The opportunity to buy it from a company that generates some of its electricity from wind farms or hydroelectric power stations is an important freedom. However, green electricity can often be more expensive than electricity generated from traditional power stations.

You could even buy your electricity from anywhere in Europe or Asia. The national grid is connected to the European grid by high voltage cables which run on the seabed of the English Channel to France.

Questions

1 Suggest ways of generating environmentally friendly green electricity.

2 Why do you think that people are prepared to pay more for electricity from a wind farm rather than a nuclear power station?

Fuel for electricity

In this item you will find out

- about the different fuels that are used to make electricity

- about the cost of using electricity

- some of the problems of making electricity with nuclear power

You can't get radiation sickness by using electricity from a nuclear power station. Neither can you get drunk on electricity made from burning alcohol – all electricity is the same, whatever fuel is used to make it!

The price you pay for your electricity depends on where you get it from. Electricity made from gas, coal, crude oil and nuclear power stations is cheap. But they all produce polluting waste, which is bad for our environment. The cost of sorting out the problems caused by these waste products in the future isn't included when you pay your electricity bill, but should it be?

▲ Cooling towers at a coal burning power station

If you buy electricity from a renewable source, such as wind power, it is relatively expensive. Of course, you aren't paying for the fuel, because there isn't any. Instead, you are paying for the research and construction of the new technology. But there is no need for future generations to pay for problems caused by pollution because wind power doesn't cause pollution.

◀ There is no polluting waste from a wind turbine

Heat energy from fuels

There are many ways of heating water to make steam in a power station:

- burning **fossil fuels**, such as coal, crude oil or natural gas
- burning **renewable biomass**, such as manure, wood from trees or straw
- splitting **uranium** atoms into smaller ones.

Burning fuels turns the chemical energy of the fuel into heat energy. When the uranium atoms in an uranium fuel rod are split, nuclear energy is changed into heat energy.

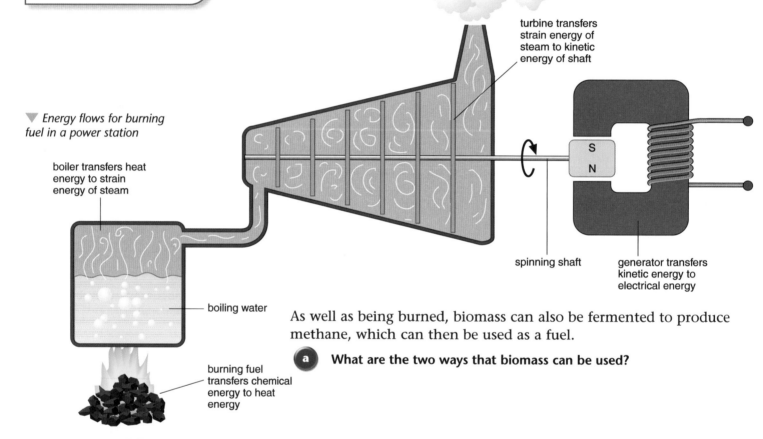

▼ *Energy flows for burning fuel in a power station*

turbine transfers strain energy of steam to kinetic energy of shaft

boiler transfers heat energy to strain energy of steam

spinning shaft

generator transfers kinetic energy to electrical energy

boiling water

burning fuel transfers chemical energy to heat energy

As well as being burned, biomass can also be fermented to produce methane, which can then be used as a fuel.

a What are the two ways that biomass can be used?

▲ *Nuclear fuel needs to be handled very carefully*

Nuclear power

Nuclear power stations use uranium as their fuel. Uranium is a non-renewable fuel and will not last for ever. One advantage of **nuclear power** is that splitting uranium atoms does not create carbon dioxide, so it does not result in global warming. But there are several disadvantages:

- nuclear waste is radioactive and gives off ionising radiation which can cause cancer
- nuclear fuel waste contains **plutonium** which is radioactive and can be used to make nuclear bombs.

b What are the advantages and disadvantages of nuclear power?

Paying for power

Electricity meters record the amount of electrical energy that passes through them, in units called **kilowatt hours** (kWh).

The cost of each kilowatt hour (or **unit**) depends on the time of day and how the electricity is generated. It can be cheaper to buy electricity at night (but only if you have Economy 7).

So how much does it cost to run an electrical heater? The cost depends on the **power** of the heater and how long it is turned on for.

First, you need to calculate the power of the heater from its current and voltage using this equation:

power = current × voltage

▲ An electricity meter shows how much electricity you use

If the heater uses a current of 12 A (amps) and a voltage of 230 V (volts) then its power in **watts** is:

power = 12 A × 230 V = 2 760 W

There are 1 000 watts in a **kilowatt**. So to convert the power in watts (W) to kilowatts (kW) you need to divide by 1 000:

$$\text{power} = \frac{2\ 760\ W}{1\ 000\ W} = 2.76\ kW$$

You need to use this equation to calculate the electrical energy used in kilowatt-hours (kWh):

energy = power × time used

If the heater is on for 6 hours every day:

energy = 2.76 kW × 6 h = 16.6 kWh

Finally, if the cost of each unit is 8 pence per kWh, then you can work out the daily cost of running the heater for 6 hours each day.

cost = 16.6 kWh × 8p per kWh = 132 pence

 An oven is on for two hours. It draws a current of 15 A from the 230 V supply. If a kilowatt hour of electricity costs 9p, show that the electricity for the oven costs 62p.

keywords

fossil fuel • kilowatt • kilowatt hour • plutonium • power • nuclear power • renewable biomass • uranium • watt

Electricity from rubbish

We throw away a lot of stuff. Cardboard, food and other items – we bin it and forget it. But rubbish can be used to make electricity.

One method is to burn the rubbish and boil water to produce high pressure steam which can be used to spin turbines to generate electricity.

But burning plastic is a tricky business – get the temperature wrong and poisonous chemicals come out of the chimney.

Fermenting the rubbish is a safer method of making electricity:

- make a big hole in the ground
- line it with a waterproof layer
- fill it with damp rubbish
- cover it with a gas proof layer
- collect the methane gas produced by microbes feeding on the rubbish
- burn the methane in gas turbines to spin generators.

This is not as expensive as it sounds. The UK has many old quarries, and they can be reclaimed and used for landfill.

Methane is even better than carbon dioxide at trapping infrared radiation, so it is better to convert it to carbon dioxide by burning it instead of letting it loose in the atmosphere to accelerate global warming.

▲ Electricity can be made from rubbish

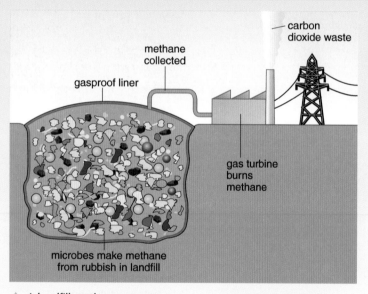

▲ A landfill methane generator

Questions

1 Explain how burning rubbish can make electricity.

2 Why is burning rubbish for electricity not a good idea?

3 Describe how to make electricity from landfill waste.

4 Why is better to burn methane from landfill waste than to let it escape?

5 Dried sewage in London is burnt to make electricity. Suggest reasons why this is more efficient than fermenting it.

Kill or cure?

In this item you will find out

- the different types of nuclear radiation
- how nuclear radiation can be used
- how to handle radioactive materials safely

Radioactivity has many uses, some of which have serious consequences for our future. Its use in medicine and industry provides enormous benefits and it can be used to make electricity, too.

But it also gives us the technology to wipe humans off the planet.

▲ *A nuclear explosion*

There is also a problem with using **radioactive materials**.

What do you do with the waste material? You cannot treat it as ordinary waste, because it emits **nuclear radiation** which is harmful. Some nuclear wastes contain low levels of radioactivity and can be buried safely in the ground in simple trenches. However, some of the waste from nuclear power stations will remain radioactive for many thousands of years.

At the moment there are three ways to deal with this problem:

- bury the waste in landfills. You can only do this if the radioactivity is low enough.
- reprocess the waste. Some of the material in the waste can be used to make fresh nuclear fuel.
- encase the most **radioactive waste** in glass blocks. These can be stored safely underground.

Amazing fact

All of the plutonium on Earth has been made in nuclear power stations. Its radioactivity will last for at least 50 000 years.

a Why is radioactive waste from nuclear power stations a problem?

b Describe how radioactive waste can be stored safely.

Nuclear radiation

Nuclear radiation is able to knock electrons out of atoms that it passes through. This **ionisation** of atoms produces charged particles. Nuclear radiation can be useful or it can be harmful. Some types of nuclear radiation can damage living cells and can cause cancer, but other forms of nuclear radiation can be used to treat cancer.

 What does the term ionisation mean?

Types of nuclear radiation

There are three types of nuclear radiation:

- **alpha** particles which can be stopped by a sheet of paper
- **beta** particles which need a few millimetres of metal to be stopped
- **gamma** rays which are partially absorbed by a few centimetres of metal.

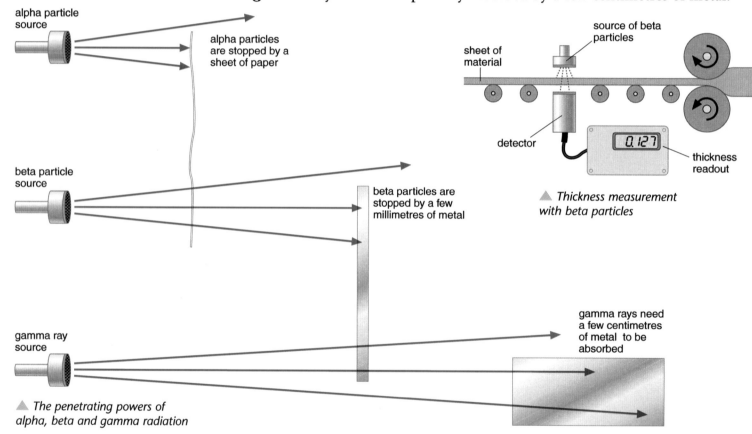

Thickness measurement with beta particles

▲ *The penetrating powers of alpha, beta and gamma radiation*

d **Name the three different types of nuclear radiation? Which is likely to be the most dangerous outside the body?**

All three types of radiation are useful. Alpha particles are used in smoke detectors. The smoke alters the ionisation of the air caused by the alpha particles.

Beta particles are used in paper thickness gauges. A source shoots beta particles from strontium-90 at a continuous sheet of paper. If the thickness of the paper increases, then the number of beta particles reaching the detector decreases.

Doctors use beams of gamma rays from cobalt-60 to kill tumours inside the bodies of cancer sufferers, avoiding the need for surgery.

Medical equipment, such as scalpels and bandages, are sterilised by gamma rays. Items are wrapped in plastic and left close to some cobalt-60 for several hours. The gamma rays penetrate deep inside the items, killing any living organism which might cause infection.

Radiation all around

You are surrounded by radiation all the time. Sources of this background radiation include:

- cosmic rays, fast moving particles from outer space
- radon, a radioactive gas given off by soil and rocks in the ground
- your food, drink and other living things.

There is no way of escaping background radiation, but the risk it poses to your health is quite small.

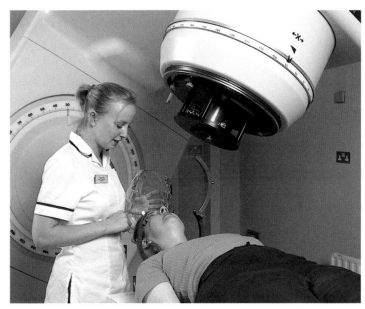
▲ Radiation can help to fight cancer

▲ Tests show that sheep from some parts of the UK are still contaminated with radioactive fallout from the explosion of the Chernobyl reactor (in Russia) in 1986

e Name three different sources of background radiation.

Handling radioactive materials

Nuclear radiation is given out by radioactive materials. Radioactive materials can cause cancer, so they need to be handled carefully:

- keep them secure in shielded and clearly labelled storage
- handle them with tongs to keep them away from you
- wear gloves and protective clothing to stop any material getting on your skin
- keep exposure times short to minimise the risk.

f Explain why radioactive materials should be locked in labelled cupboards.

▲ Radioactive material must carry a hazard symbol

keywords

alpha • background radiation • beta • cosmic ray • gamma • ionisation • nuclear radiation • radioactive materials • radioactive waste

Scanning bodies

One of Steve's testicles developed cancer twenty years ago. The testicle was removed in time, and he is OK. However, there is a small chance that some cancer cells got away. So Steve has a body scan every few years to check that he isn't getting cancer anywhere else.

When Steve arrives at the hospital, Tracy injects him with a radioactive liquid. 'It emits gamma rays,' says Tracy. 'Since I handle the stuff every day at work I need to be careful with it. I don't want to get cancer myself!'

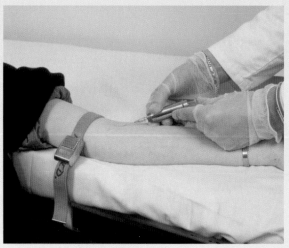
▲ The radioactive material is injected

Steve quickly gets into the gamma camera. This picks up gamma rays from all parts of his body at once. As soon as he is comfortable, Tracy leaves him. This reduces her exposure to gamma rays.

The radioactive liquid is carried all over Steve's body in his blood. It collects in places where the cells are dividing rapidly, such as cancer tumours. So these places emit more gamma rays than others. The doctors in the next room can see this from their display. After an hour lying in the gamma camera, Steve gets the good news – no sign of cancer this time.

'It's weird,' says Steve. 'They make me radioactive, so that could give me cancer. Yet it could save my life if I get cancer again. Anyway, the radioactivity only lasts a few days, so I'm not a health risk for very long!'

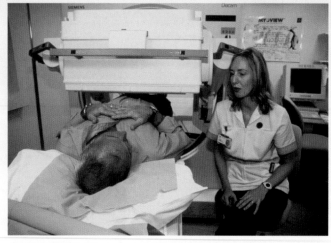
▲ Steve lies in the gamma camera

▲ The detected gamma rays build up an image

Questions

1 Explain why the syringe is inside a thick tube of lead.

2 Describe the safety precautions taken by Tracy.

3 Why does the injected liquid have to emit gamma rays? Why not alpha particles or beta particles?

4 Explain how the camera detects possible cancer tumours.

The cosmic ray shield

In this item you will find out

- about the Earth's magnetic field

- how the Moon may have been formed

- that solar flares can damage satellites orbiting the Earth

Three thousand years ago, the Chinese invented the magnetic compass. They floated an iron-rich rock, called magnetite, on water and found that it always pointed in the same direction. It was to become a crucial instrument for safe navigation at sea.

▲ *A ship's compass*

We now know that a magnetic compass indicates the direction of the Earth's **magnetic field**. This field surrounds the Earth and shields us from the worst effects of cosmic rays – the deadly ionising radiation from the Sun and outer space.

But where does the Earth's magnetic field come from? And why is it so strong? The magnetic field around similar-sized planets is much weaker. One theory suggests the collision of two large planets. All of the heavy iron ends up in the larger planet, allowing it to have a strong magnetic field and be safe for living organisms. A lot of evidence supports this as the origin of our own Earth-Moon system. Perhaps life on Earth is only possible because of a catastrophic collision a long time ago.

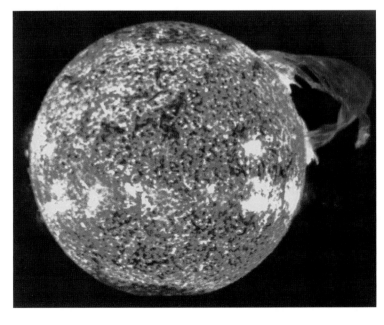

◀ *Solar flares are one source of cosmic rays*

Magnetic field

Earth is battered on all sides by particles from outer space. They are called cosmic rays. They ionise materials that they pass through, so are bad news for living organisms. Cosmic rays cause cancer. Fortunately for life on Earth, a lot of cosmic rays are made harmless by the Earth's magnetic field.

Cosmic rays are charged particles. They change direction as they pass through a magnetic field. So the Earth's magnetic field acts a shield against fast moving particles.

a   **What are cosmic rays? Why are they a bad thing?**

The diagram shows the shape of the Earth's magnetic field. Like a magnet, the magnetic field has a North Pole and a South Pole.

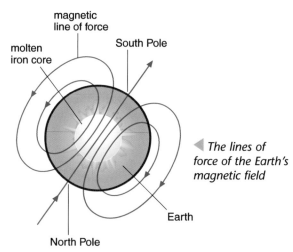

The lines of force of the Earth's magnetic field

Where does the Earth's field come from? Scientists think that it is made by large electric currents set up in the liquid core of iron as the Earth spins on its axis. The shape of the field is very like the one around a coil of wire when it carries a current.

b   **Explain how scientists think the Earth's magnetic field is created.**

Making the Moon

Our planet has a very large molten iron core compared with other moons and planets visited by our spacecraft. Without that iron, we would have no magnetic field. The Moon does not have an iron core. Scientists think that a catastrophic collision of two large planets, which may have happened four billion years ago, could explain this.

A planet the size of Mars may have collided with Earth. All of the heavy iron from the cores of both planets merged to create the Earth while all the lighter material merged to form the Moon.

Astronauts have brought rocks back from the Moon. They are very similar to rocks on the surface of the Earth, suggesting that both do indeed have the same source.

c   **Suggest why the Earth has such a large magnetic field.**

d   **Explain how the Moon may have been formed.**

An electrical current (moving electrical charges) passing through a coil creates a magnetic field

Examiner's tip

Don't confuse explanation with description.

Amazing fact

A magnetic compass is a small magnetised steel needle free to rotate about its centre. The needle always points in the direction of the local magnetic field.

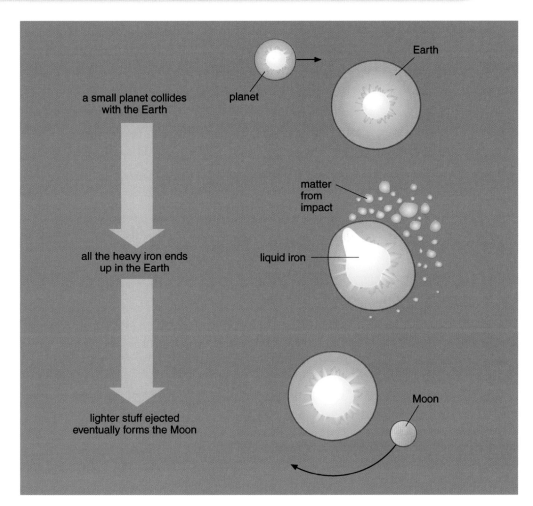

▲ *How the Moon may have been formed*

Satellites and solar flares

We have put lots of **satellites** in orbit around the Earth to serve different functions:

- communications satellites use microwaves to pass information from one place on Earth to another
- weather satellites send back pictures of cloud formations, allowing accurate weather forecasts
- spy satellites monitor activities in other countries
- navigation satellites send out GPS (global positioning system) signals.

Many communication satellites orbit beyond the Earth's magnetic field. This means they can be badly damaged by **solar flares**. These are vast clouds of charged particles ejected at high speed from the Sun from time to time. They produce strong magnetic fields.

Particles from solar flares are similar to cosmic rays and cause ionisation, so they can also cause cancer.

 Why can solar flares interfere with our communications systems?

keywords

magnetic field • solar flare
• satellite

Blackout

▲ *Solar flares can turn all these off in a matter of seconds*

On 13 March 1989, Quebec's power supply failed completely, leaving millions of people without electricity for days.

The damage was caused by the Sun. A huge cloud of swirling charged particles had been launched in Earth's direction the previous day, travelling at over a million kilometres per hour.

When it slammed into the Earth's magnetic field over Canada, it was like two giant magnets colliding.

The resulting surges of magnetism on the Earth's surface created pulses of electricity in large metal loops such as those formed by transmission line systems.

There was enough energy in those pulses to burn out transformers, bringing the electricity distribution system to a sudden stop.

We may be able to avoid this sort of damage in the future. We have placed satellites in orbit around the Sun to look out for solar flares heading straight towards the Earth. The flares take a few hours to make the journey, giving power station operators time to prepare. They don't need to switch off the power, just reconnect the transmission lines so that they don't form large loops which can generate electricity when the change of magnetism happens.

▲ *Surges of magnetism can cause large voltages*

Questions

1 Explain how the Sun caused the power failure in Quebec.

2 Suggest how we could avoid this sort of power failure in the future.

3 What else might be damaged by solar flares hitting the Earth?

4 Suggest why optical fibre networks on Earth are not affected by solar flares.

Is anybody there?

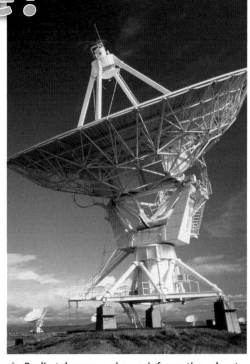

In this item you will find out

- about the different objects in the Universe

- about the difficulties of manned space travel between planets

- how robot spacecraft allow us to explore the Solar System

One of the biggest unanswered questions is, 'Are we alone?' We know that there is life on Earth, but what about elsewhere in the Universe?

Humans have always been interested in the stars and watched for intelligent communication from them. We watched with naked eyes at first, then with optical telescopes and finally across the entire electromagnetic spectrum for messages from aliens. As yet, there is no scientific evidence of anything out there trying to communicate with us. But we keep on listening …

▲ *Radio telescopes give us information about the Universe*

Of course, life may be everywhere in the Universe, but intelligence capable of beaming signals across the galaxy may be rare. After all, life on Earth began with bacteria, as soon as it was cool enough for water to stay liquid.

That was about 4 billion years ago. Our own species, *Homo sapiens* has only been around for about 4 million years, and has only acquired the technology to beam out radio signals in the last 40 years.

But the search is on for life of any sort in our own Solar System. Robot spacecraft are visiting the most likely habitats, and evidence is beginning to suggest there might be bacteria on other planets.

▲ *A Mars spacecraft*

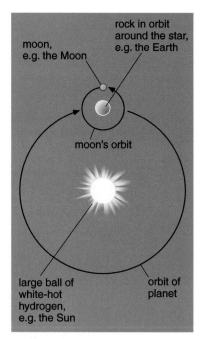

The Moon orbits the Earth which orbits the Sun

The Universe

We inhabit the surface of a **planet**, one of nine in **orbit** around a large ball of white-hot hydrogen gas. This is the **star** that we call the Sun. Like all stars, it is so hot that it emits huge quantities of light. There is a lump of rock which orbits around our planet. We call it the Moon. The Earth has only one moon, but other planets have many more. Some have none. The Universe also contains **comets** and **meteors**, **black holes** and groups of stars called **galaxies**.

The Sun is the largest object in the Solar System and its gravitational attraction keeps the smaller planets in orbit around it. For the same reason, moons are smaller than the planets that they orbit. Most of the planets have near-circular orbits in the same plane. Pluto does not, which suggests that it was captured from outer space after the other planets had been created. Gravitational force determines the motion of planets and satellites.

Planet	Orbit radius (light-minutes)	Orbit time (years)
Mercury	3.2	0.24
Venus	6.0	0.62
Earth	8.3	1.0
Mars	13	1.9
Jupiter	43	12
Saturn	79	29
Uranus	160	84
Neptune	250	165
Pluto	330	250

a How does the orbit time change as the orbit radius increases?

The size of the orbit is given in light-minutes. This is the distance that light travels in one minute and is 18 000 000 km.

b Show that a radio signal would take over 5 hours to get from Pluto to Earth.

Amazing fact

The nearest star is over 4 light-years away. We can see it clearly in the night sky because it is very hot and gives off lots of light.

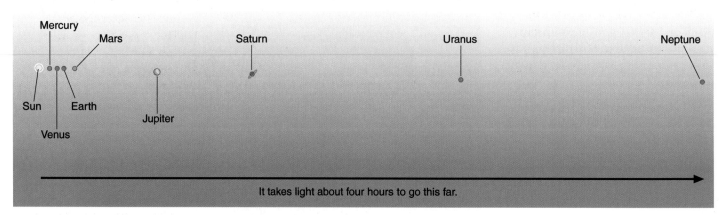

The relative positions of the planets to the Sun

This spiral galaxy contains billions of stars

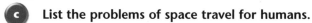

photocells make electricity from sunlight

shiny surface to reduce radiation

metal foam to reduce conduction

spherical shape reduces surface area for radiation

stores

living quarters at +20°C

space at −270°C

Manned space travel

It is going to be difficult to get human beings to other planets:

- the voyage will take a long time, perhaps many years
- a lot of energy will be needed for heating and maintaining a stable atmosphere
- getting away from the Earth's gravity requires a lot of fuel
- cosmic rays and solar flares will dramatically increase the risk of cancer, so astronauts will need to be shielded and protected from them
- astronauts will need a supply of food, fresh air and clean water for the length of the journey.

c List the problems of space travel for humans.

Robots in space

Unmanned **spacecraft**, such as probes, have done most of the exploring of our Solar System. Onboard computers control the spacecraft, following commands sent by radio from Earth. Computers don't need food air or water, so robot spacecraft can easily be launched with rockets. They can also explore areas where conditions are lethal to humans. Their cameras have given us stunning images of other planets, using infrared and ultraviolet as well as visible light. Other sensors on the spacecraft can send back information about the temperature, magnetism, gravity and atmosphere on other planets.

d What are the advantages of exploring the Solar System with robots instead of humans?

▲ *Mars taken from an orbiting spacecraft*

Amazing fact

Some of the first robot spacecraft have already left the Solar System. They won't arrive at the nearest stars for thousands of years.

keywords

black hole • comet • galaxy • meteor • planet • orbit • spacecraft • star

Life in low gravity

NASA plans to send a manned mission to Mars in about 2025. Most of the technology required exists today. The real problem is the payload of the mission, which consists of the people it transports.

Humans are poorly adapted to the environment of space.

▲ *Astronauts lose bone mass without the stress of walking, running and jumping*

This is because our bones change in response to their use. The continual stress of walking, running and jumping makes them thicker and stronger, but lack of stress makes them thinner and weaker. This is why astronauts on the Mir Space Station lost about 2% of their bone mass each month, even though they had an exercise programme to stress their bones.

Any astronaut on a journey to Mars will spend a year in an environment with almost no gravity. They will lose 30% of their bone mass by the time they arrive, dramatically increasing their chances of breaking a leg when they walk on the surface.

The solution may be to set the spacecraft spinning. Astronauts inside the spacecraft will need a force from the hull to keep them moving in a circle. Get the speed right, and this force will mimic Earth's gravity.

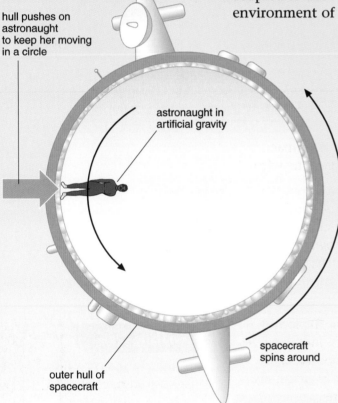

hull pushes on astronaught to keep her moving in a circle

astronaught in artificial gravity

outer hull of spacecraft

spacecraft spins around

▲ *A spinning spacecraft mimics the Earth's gravity*

Questions

1 What happens to bones in low gravity?

2 Explain two ways of getting around the problem.

3 Increasing (or decreasing) the speed of the spacecraft by 10 m/s every second will make the astronauts feel as if they are in Earth's gravity. Suggest why this isn't possible for a long space mission.

4 Once the spacecraft is spinning, no fuel will be needed to keep it going. Suggest why.

Catastrophe!

In this item you will find out

- the difference between an asteroid and a comet

- what would happen if an asteroid hit the Earth

- about Near-Earth Object observations

▲ *A meteor crater in Arizona, USA*

If you look at the Moon when it is full, it looks perfectly smooth and round. The first person to look at the Moon through a telescope was Galileo. He immediately noticed the roughness of the surface, with its flat plains ringed by tall mountains.

Each of the circular features on the Moon's surface looks like an impact crater. It is what you get when a small object hits a much larger one with enough energy to melt part of it.

The Moon is covered with many craters – large ones, small ones and overlapping ones. Scientists use this evidence of violent impacts to support their ideas of planet formation. Gravity tugs dust and rocks towards each other, so that they collide and stick. As objects grow in this way, the impacts get more and more spectacular, until all the loose matter has been swallowed up into a moon or a planet.

With no wind, frost or rain to change its surface, the record of those violent times is preserved for ever on the surface of the Moon. But the violent times may not be over. There are craters on Earth which look just like those on the Moon. They can't be old, or wind and rain would have worn them away by now. So perhaps there are still large lumps of rock out there ...

▲ *Galileo was the first person to look at the Moon's surface in detail*

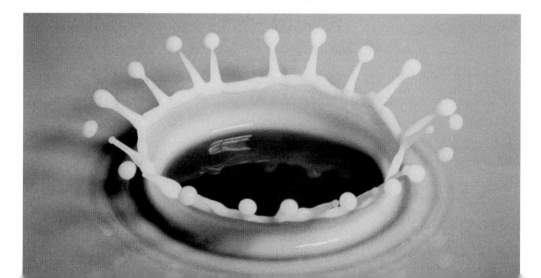

◄ *How a crater is formed*

▲ *An asteroid travelling towards Earth*

Impact

Asteroids are large rocks left over from the creation of the Solar System. Lots of asteroids are trapped by the gravity of Mars and Jupiter in the Asteroid Belt. But collisions can push them out of a safe, circular orbit well away from the Earth, into an ellipitical orbit which crosses Earth's orbit.

 a Why are asteroids safe in circular orbits but unsafe in elliptical ones?

The Earth moves through space at about 110 000 kilometres per hour as it orbits the Sun. What would happen if an asteroid got in the way?

- the kinetic energy of the asteroid would be the same as a million atom bombs
- the asteroid would melt and vapourise as it plunged many kilometres into the Earth's crust
- this explosive transfer of kinetic energy into heat would fling vast quantities of dust and debris into the air
- a ring of molten rock would be flung up around the impact point, making a crater
- white hot material ejected into space by the impact would fall back to Earth over a wide area, causing widespread fires
- the dust would linger in the atmosphere for years, blocking out the sunlight which feeds the plants
- the change in climate would lead to many organisms dying from lack of food.

b How could an impact on one part of the Earth affect the well-being of animals everywhere?

Scientists have found evidence that asteroids have collided several times with the Earth in the past. The last time was 65 million years ago, possibly causing the extinction of the dinosaurs. Scientists have found impact craters and rock layers with unusual elements. Scientists have also found sudden changes in the numbers of fossils in layers of rock lying next to each other.

▲ *Optical telescopes allow astronomers to track NEOs*

Looking out for NEOs

A **Near-Earth Object** (NEO) is anything big out there, such as an asteroid or a comet, which might collide with the Earth.

Both an asteroid or a comet could cause a lot of damage if they hit us. Astronomers track them with telescopes and calculate the chances of a collision in the future. But the relatively small mass of an NEO means that its orbit is easily changed by other planets it passes close to!

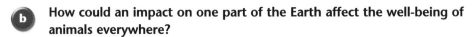

c What could we do to avoid being hit by an NEO?

Comets

Comets are made from material orbiting the Sun from beyond the planets. This material consists of cold chunks of ice and dust in large slow orbits beyond Pluto. The ice and dust can be jolted into highly **elliptical** orbits around the Sun by the gravity of passing stars.

A comet spends most of its time far away from the Sun. As it approaches the Sun the stronger gravity makes it speed up. Heat from the Sun melts the ice, and it sheds material to form a tail.

A NEAR-EARTH OBJECT THE ASTRONOMERS DIDN'T SPOT

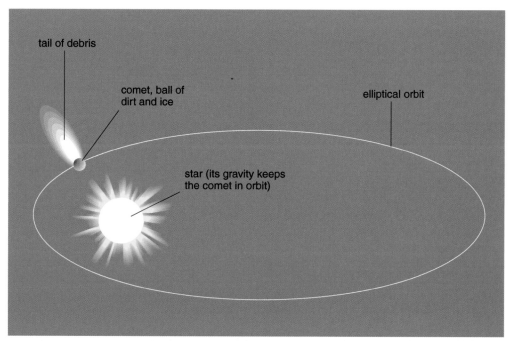

tail of debris

comet, ball of dirt and ice

elliptical orbit

star (its gravity keeps the comet in orbit)

▲ The elliptical orbit of a comet

d **Comets don't last for ever. Why not?**

▲ Halley's comet in 2002

keywords

asteroid • crater • elliptical • Near-Earth Object

Near miss in 2002

On Sunday 16 June 2002, most people were preoccupied with football. Who would win the World Cup in Japan? Nobody noticed that an asteroid the size of a football pitch missed Earth by a mere 75 000 miles. This is only three times the Earth's diameter, less than the distance between us and the Moon.

When the asteroid was spotted three days later, it was christened 2002MN. It had a diameter of between 60–120 m.

Had it hit the Earth, nothing would have survived within a distance of 1 000 miles of the impact point. An impact in a heavily populated area, such as England, would have resulted in a huge loss of life.

According to the Near-Earth Object information centre based in Leicester, the last time we had such a close shave was in December 1994.

So why was 2002MN not spotted before it passed Earth? Astronomers can only use their telescopes at night, when the Sun is blocked by the Earth. They cannot see anything approaching the Earth from the direction of the Sun.

At present, the only government funded agency to keep track of NEOs is NASA in the USA.

It only tracks objects that are bigger than 1 000 m across, using telescopes in the northern hemisphere. There is no government programme to look for NEOs approaching the southern hemisphere.

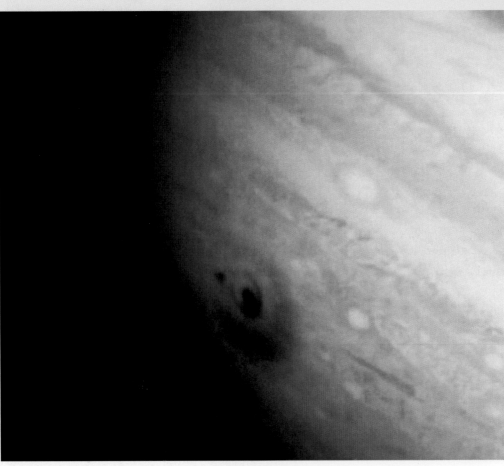

▲ *Jupiter just after it was hit by pieces of a broken comet. The black spots show the points of impact. Each is about the size of Earth*

Questions

1 What is 2002MN?

2 Explain how an object only 100 m across can damage an area over 1 000 000 m across.

3 Suggest why nobody from the southern hemisphere is looking for NEOs.

4 Explain why 2002MN was only spotted after it had passed the Earth.

Life and death of stars

In this item you will find out

- about the possible origins of the Universe
- that the Universe is expanding
- what happens when stars die

Have you ever looked up at the night sky and wondered how long it has been there?

Our knowledge of the Universe comes from interpreting its electromagnetic waves. Telescopes allow us to survey the sky over the whole electromagnetic spectrum, from radio waves to gamma rays.

All the waves tell us one story – that the Universe appeared explosively out of nothing 15 billion years ago.

The key to this story was discovered by accident. In 1814, Joseph von Fraunhofer noticed that when a glass prism split sunlight into colours, there were black bands in the spectrum.

In 1823 he tried the same experiment with starlight and discovered that the black bands were in different places. They were shifted to the red end of the spectrum and they had a longer wavelength.

In 1842, Christian Doppler put the last piece in the jigsaw by discovering that the wavelength of a wave increases as you move away from its source.

So, stars with red-shifted light must be moving away from us. The red shift of starlight means that the Universe is expanding!

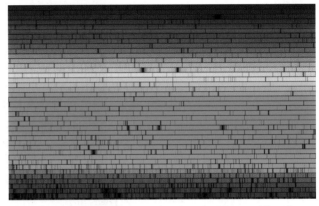

▲ A spectrum of the Sun – note the black vertical bands

▲ White light is made up of different colours

 a　The light from the star Zarg is shifted towards the blue end of the spectrum. What does this tell you about Zarg?

The expanding Universe

Stars are clumped together in galaxies, with about a billion stars in each. Scientists can measure the speed of these galaxies by looking at the red shift of the light from their stars. These measurements provide evidence that:

- almost every galaxy is moving away from ours
- the further away a galaxy is, the faster it moves.

A galaxy of stars ▶

 How does the red shift of starlight from galaxies suggest that the Universe is expanding?

The origin of the Universe

The red shift of galaxies provides evidence for a theory of the origin of the Universe. It is called the **Big Bang theory**:

- a very hot Universe appeared 15 billion years ago at a single point
- it expanded rapidly and cooled at the same time
- it is still expanding.

The Universe is filled with microwave radiation. This is because the radiation left over from shortly after the Big Bang has been stretched from very short wavelength gamma radiation to long microwave radiation (centimetres long) as the Universe expanded.

 What is the Big Bang theory?

Amazing fact

Our Sun has enough hydrogen fuel to keep going for the next 5 billion years.

Star birth

The Universe is full of hydrogen gas. How does the gas become a star? The pattern is as follows:

- it starts off with a large cloud of hydrogen
- gravity tugs every atom to every other atom
- so the cloud shrinks because of gravity
- gravitational energy is transferred to heat energy
- until fusion reactions start at the hot centre of the cloud
- releasing energy to stop the shrinking.

The fusion reactions start at a temperature of about 15 million °C, and convert hydrogen into helium. The energy released is in the form of heat and light.

d   **Explain how a large cloud of hydrogen becomes a star.**

▲ *A star field in outer space*

Star death

When a star like our Sun starts to run out of hydrogen, it becomes unstable and swells up to become a **red giant**. Once the fusion reactions have stopped, the star cools and shrinks, becoming a **white dwarf**. When it is so cool that it can no longer emit light, it becomes a **black dwarf**.

Stars which start off about ten times larger than our Sun have a different fate. They are so hot that they appear blue. As the hydrogen runs out, they expand into a red giant, but fusion reactions only stop when all the material deep inside has been converted to iron.

The star then explodes as a **supernova**, scattering most of its material into space, where it can eventually form other stars and the planets which orbit them. A tiny **neutron star** is left behind, made of superdense matter.

If the star is heavy enough, the supernova leaves behind a black hole, so massive that not even light can escape from it.

e   **Describe what happens to a star when it runs out of hydrogen fuel.**

keywords

Big Bang theory • black hole • neutron star • red giant • supernova • white dwarf • black dwarf

Unstable Universe

Less than a hundred years ago, scientists believed that the Universe just consisted of a large number of stars scattered through space. They knew that Space was big, but not how big.

All the evidence pointed to a Universe that didn't seem to change – a Steady State Universe. So they proposed the Steady State theory – that hydrogen atoms appeared in space as fast as they were used up in stars. You would only need one new atom per cubic kilometre of space in every ten years – an idea that is very difficult to disprove.

Scientists began to work out the consequences of Einstein's General Theory of Relativity when it was first published in 1917. This showed that a Steady State Universe was very unlikely. The Universe should either be expanding or collapsing. Even Einstein wasn't very happy with this.

▲ *Large reflecting telescopes make investigation of the stars possible*

It was only a few years later that Edwin Hubble found evidence to support the idea that the Universe is expanding. He used the biggest telescope of his time, with a reflecting dish of 2.5 m diameter, to peer deeper into space than anyone else.

Not only did he discover that stars are actually clustered into galaxies, he showed that their light was red-shifted. The Big Bang theory was born out of the idea that this was due to the expansion of the whole Universe.

The Big Bang theory eventually replaced the Steady State theory because it was used to make a number of predictions, which were later supported by other evidence gained from modern telescopes.

These include:

- the observed ratio of hydrogen to helium in the Universe
- the presence of microwave radiation from everywhere
- stars from the early Universe, with a very large red shift, can be quite different from the ones around us today.

Questions

1 What does the Steady State theory say?

2 What was the evidence for the Steady State theory when it was proposed?

3 Give three pieces of evidence which the Steady State theory could not explain.

4 Suggest why Einstein's ideas were important for the Big Bang theory.

P2a

1 Complete the sentences. Choose from:

electricity   heat   light   magnetism   sound

The Sun transfers energy to Earth as ____(1) and ____(2).

It can be transferred into ____(3) by photocells.   [3]

2 The sentences explain one way of using the Sun to make electricity. Put them in the correct order.

A to make a convection current
B the heat is transferred to the air
C light energy comes from the Sun
D the ground transfers it to heat energy
E which makes wind turbines spin round   [4]

3 Photocells can be used to make electricity.

State two advantages and two disadvantages of using photocells.   [4]

4 Describe three different ways in which energy from the Sun can be used to provide energy for our homes.   [3]

5 Some people think that wind turbines are best placed out at sea.

What are the advantages of this? What are the disadvantages?   [4]

6 A solar heater has water pumped through pipes exposed to the Sun. The pipes are kept under glass and coloured black. The sentences explain how a solar heater works.

Put them in the correct order.

A Sunlight passes through the glass.
B The infrared is trapped by the glass.
C The pipes heat up emitting infrared radiation.
D The water carries the heat energy into the house.
E Light energy is absorbed by the surface of the pipes.
F Heat energy conducts through the pipes to the water.   [4]

P2b

1 Match the start and end of each sentence.

1 a consumer takes   a direct current
2 a battery produces   b alternating current
3 a power station puts   c electricity into the grid
4 a generator produces   d electricity from the grid   [3]

2 While a magnet is being placed in a coil, the voltmeter connected to the coil reads +5 V.

a What does it read when the magnet is left in the coil?
b What does it read when the magnet is removed from the coil?   [1]

3 What does a transformer do?   [1]

4 A generator has a coil of wire next to a spinning magnet.

State three ways of increasing the voltage from the generator.   [3]

5 The sentences describe the energy transfers in a power station. Put them in the correct order.

A which makes the generator spin round
B which passes through the turbine
C water is boiled to make steam
D the fuel is burnt   [3]

6 Explain how transformers are used in the national grid.   [4]

7 Power stations produce a lot of waste heat, in the form of water at about 90°C.

Suggest two ways of putting that heat to use, and explain why it isn't already happening in the UK.   [4]

P2c

1 Name three fossil fuels.   [3]

2 Name two types of biomass which can be used to make electricity.   [2]

3 Complete the sentences. Choose from:

**current   length   power   speed
time   voltage**

The cost of running an electrical appliance depends on its ____(1) and on the amount of ____(2) that it is running.   [2]

4 State one advantage and one disadvantage of using nuclear power to make electricity.   [2]

5 A kettle has a current of 9 A when connected to a 230 V supply. Use the equation

power = voltage × current

to calculate the power of the kettle.   [2]

6 Calculate the cost of running a 2 kW heater for 6 hours when a unit of electricity costs 8p.   [3]

7 Match the start and end of the sentences.

1 methane is a product   a of burning a fuel
2 plutonium is a waste   b of a nuclear reactor
   product
3 heat energy is   c of fermenting biomass
   transferred as a result   [2]

8 A power station generates electricity by burning sawdust from a sawmill.

Explain why the waste gases from this power station do not contribute to global warming, even though they are made from carbon dioxide. [2]

P2d

1 Name the three types of nuclear radiation. [3]

2 Describe one useful application of nuclear radiation. [1]

3 State three precautions to be taken when handling radioactive materials. [3]

4 **a** What is background radiation? [1]
b Where does it come from? [2]

5 Describe one use each for alpha particles, beta particles and gamma rays. [3]

6 Describe two ways of disposing of radioactive materials [2]

7 Nuclear radiation ionises materials. Explain what this means. [2]

8 Describe how to safely dispose of waste materials that are highly radioactive.

Why are these substances such a problem? [3]

P2e

1 Which one of these statements about the Moon is most likely to be correct?

A the Moon exploded out of the Earth billions of years ago
B the Moon is the remains of a planet which hit the Earth
C the Moon is the remains of a star which has run out of fuel [1]

2 State three uses of artificial satellites around the Earth. [3]

3 Match the start and end of the sentences.

1 a magnet has	a makes a magnetic field
2 there is a magnetic field	b around the Earth
3 a compass can be used	c a north and a south pole
4 a current in a coil	d to find the direction of the field [3]

4 Describe how a collision between two planets can give rise to an Earth-Moon system. [3]

5 The Earth's magnetic field protects us from solar flares.

a What is a solar flare? [2]
b How does the magnetic field protect us? [1]

6 Complete the sentences. Choose from:

**cells   charged   magnetic   poles
Sun   surface**

Cosmic rays are mostly ____(1) particles from the ____(2).

They are deflected by the ____(3) field around the Earth, so not many of them get to its ____(4) where they can damage living ____(5).

This protection does not exist at the North and South ____(6). [6]

7 Here are some statements about the Moon. Which of them are evidence to support the idea that the Moon was created from a collision between a planet and the Earth?

A There is no air on the Moon.
B The Moon has no magnetic field.
C The Earth has a large magnetic field.
D The Moon is smaller than the Pacific Ocean.
E The surface of the Moon is covered in craters.
F Moon rocks brought back by astronauts are similar to rocks on the Earth's surface. [3]

P2f

1 List these items in order of increasing size:

**comet   galaxy   meteor   planet
star** [4]

2 Which one of these things are needed by unmanned spacecraft?

A air
B electricity
C food
D water [1]

3 Complete the sentences. Choose from:

Earth   Moon   Sun.

The ____(1) orbits the Earth.

The Earth orbits the ____(2). [2]

4 Providing enough food and water will be difficult for a manned voyage between planets.

State four other difficulties. [4]

5 Name the third and fifth planets from the Sun. [2]

6 Unmanned spacecraft have sent back information about other planets.

State four different things they have told us. [4]

7 The sentences explain how a planet's distance from the Sun affects its temperature. Put them in the correct order.

A They spread out in straight lines as they travel through space.
B A far off planet only intercepts a few of these rays.
C A nearby planet intercepts lots of these rays.
D So it absorbs lots of energy from the Sun.
E So it absorbs less anergy from the Sun.
F Light rays are emitted from the Sun. [5]

P2g

1 Complete the sentences. Choose from:

**bounced   collided   dust   ice   Jupiter
Mars   orbited   rocks   Venus**

Asteroids are ____(1) some of which have ____(2) with Earth in the past.

They normally orbit between ____(3) and ____(4). [4]

2 The sentences describe what happens when an asteroid impacts on Earth. Put them in the correct order.

A starting fires where they land
B many plants and animals die
C sunlight is blocked by dust
D hot rocks are thrown up
E a crater is formed [4]

3 Match the start and end of the sentences.

1 a comet is   a made from rock
2 an asteroid is   b a comet or asteroid
3 a Near-Earth Object is   c made from dust and ice
[2]

4 There is evidence on Earth for collisions with asteroids in the past.

Describe three pieces of the evidence. [3]

5 A comet recently hit Jupiter.

a Where do comets come from? [1]
b What are comets made of? [1]
c Why does a comet have a tail? [2]

6 How do scientists know if a NEO is going to impact on Earth? [2]

7 Suggest two things that we could do to a NEO to stop it hitting the Earth. [2]

P2h

1 The sentences are about the Big Bang theory. Put them in the correct order.

A a huge explosion
B and then continues to expand
C the Universe comes into being [2]

2 Match the start and end of the sentences.

1 a black hole   a collapses to make a star
2 a gas cloud   b doesn't emit light for ever
3 a star   c doesn't allow light to escape [2]

3 The sentences are about the story of the Sun. Put them in the correct order.

A stopping the collapse by emitting heat and light
B then collapses and cools to form a white dwarf
C when hydrogen runs out, the star swells to a red giant
D a cloud of hydrogen gas collapses under its own gravity
E heating up until fusion reactions start to convert hydrogen to helium [4]

4 Describe the end of a large star. [3]

5 How is the speed of a galaxy related to its distance? [3]

6 What two things does the red shift of the light from a galaxy tell us about that galaxy? Choose from

**age   colour   distance   gravity   shape
size   speed** [2]

7 Match the start and end of the sentences.

1 a red giant is   a a star which has stopped fusion reactions
2 a planetary nebula is   b an exploding star
3 a white dwarf is   c a star which has run out of hydrogen fuel
4 a black hole is   d hydrogen fusing to make helium
5 a star is   e the massive remains of a supernova
6 a supernova   f a cooling star which sheds material into space [4]

Can-do tasks

Do you like doing practical things and getting credit for what you can do? If your answer is yes, then you will like Can-do tasks.

There are 81 of these tasks throughout your GCSE Science course. Some are practical and some require the use of ICT.

You can only count a maximum of eight of these tasks towards your final mark and the tasks are set at three levels.

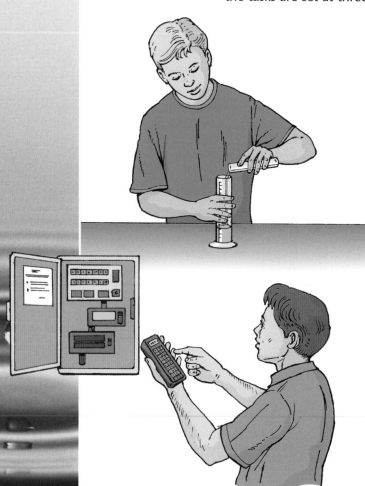

Level 1 (worth 1 mark): These are simple tasks that you can usually complete quickly. You may have done many of these before you started the course.

Level 2 (worth 2 marks): These are slightly harder tasks that may take a little longer to do.

Level 3 (worth 3 marks): These are even more difficult tasks that may take you some time to do.

Here are three examples, one from each level:

I can measure the volume of a liquid using a measuring cylinder.

This is a Level 1 task. You will do this many times during the course.

I can use meter readings to calculate the cost of using electricity.

This is a more difficult task in which you have to complete some calculations. It is a Level 2 task. Don't worry if you don't like calculations. There are many other tasks for you to try.

I can use ICT to produce an information leaflet on one endangered species showing reasons for its predicament and suggestions for its protection.

This will take some research and some time to do. This is a Level 3 task.

Your teacher has to see that you have completed a task and record this on a record sheet.

Your teacher may give you a list of all the Can-do tasks at the start of the course. You can tell him or her when you think you have successfully completed one of them. They can then give you credit for this.

Remember, it does not matter if you fail to do a task or if you are absent. There will be many more chances throughout the course.

Finally, just how important is a Can-do task? Every time you complete even a Level 1 task it is worth more to your final result than scoring a mark on a written paper.

Good luck!

Science in the News

Do you read a newspaper or listen to radio or television news programmes?

Do you believe everything you read or hear?

Two headlines in a national newspaper on 28 October 2005 were:

'Bird Flu Man is a Hospital Worker'

and

'Weekly Helping of Broccoli May Cut Lung Cancer Risk'

Looking at the first headline you might think the man had bird flu. If you had read the article carefully you would have discovered that he was the man who ran a quarantine centre where two parrots died and he worked in a hospital. He never had bird flu.

Reading the second article would tell you that the sample used in the study was so small that scientists could not be sure that a weekly helping of broccoli would really cut the risk of getting cancer.

As part of your GCSE core Science you have to do at least one Science in the News task. If you do more than one, your best mark will count. The task will be in the form of a question. For example: Should smoking be banned in public places? With the question you will be given some 'stimulus' material to help you and about one week to do some research.

What should you do with the stimulus material?

Read it through carefully and identify any scientific words you do not understand. Look up the meaning of these words. The glossary in this book might be the starting point. Then go through with a highlighter pen and highlight those parts of the stimulus material you might want to use to answer the question.

What research should you do?

You should be looking for two or three sources of information. These could be from books, magazines, the Internet, or CD-ROMs. You could also use surveys or experiments.

You will need to include with your report a list of sources that are detailed enough that somebody could check them. Some of your sources could look like this:

1 www.webelements.com/webelements/scholar/index.html

2 Heinemann, *Gateway Science: OCR Science for GCSE*, Higher text book (2006) p. 75–77

3 *Daily Mail*, 26th October 2006 p.2.

You can take this research material with you when you have to write your report about one week later. Do not print out vast amounts of material from the Internet because you will not be able to find what you want when you write your report. Your teacher will probably collect in your research material to help them to assess your report, but they will not actually mark it.

If you choose to do no research it does not stop you writing a report but you will get a lower mark.

Writing your report

You will have to write your report in a lesson supervised by the teacher. It has to be your own work.

Your report should be between 400 and 800 words. As you write your report, you need to refer to the information you have collected. For example, 'scientists believe that the long-term demand for electricity in the UK can only be met with a new generation of nuclear reactors (3)'. The 3 in brackets refers to source 3 from your list of sources, *Daily Mail*.

You should be critical of the sources. You will soon find that everything you read in newspapers is not necessarily true. Make sure you answer the question. When you finish, read your report through carefully.

Marking your report

Your teacher will mark your report and look for six skills, each marked out of a maximum of six. This makes a total of 36.
Your report is worth about 20% of the total marks. Hopefully, writing this report will make you more aware of science in our everyday lives.

Useful data

Physical quantities and units

Physical quantity	Unit(s)
length	metre (m); kilometre (km); centimetre (cm); millimetre (mm)
mass	kilogram (kg); gram (g); milligram (mg); micrograms (µg)
time	second (s); millisecond (ms)
temperature	degree Celsius (°C); kelvin (K)
current	ampere (A); milliampere (mA)
voltage	volt (V); millivolt (mV)
area	cm^2; m^2
volume	cm^3; dm^3; m^3; litre (l); millilitre (ml)
density	kg/m^3; g/cm^3
force	newton (N)
speed	m/s; km/h
energy	joule (J); kilojoule (kJ); megajoule (MJ)
power	watt (W); kilowatt (kW); megawatt (MW)
frequency	hertz (Hz); kilohertz (kHz)
gravitational field strength	N/kg
radioactivity	becquerel (Bq)
acceleration	m/s^2; km/h^2

Electrical symbols

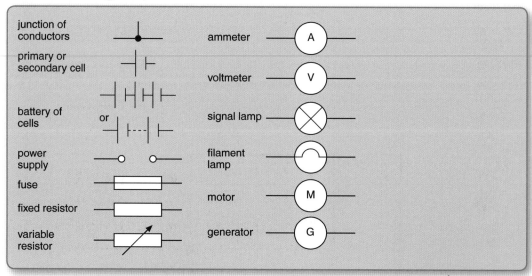

Periodic Table

Key

	relative atomic mass
	atomic symbol
	name
	atomic (proton) number

1	2												3	4	5	6	7	8
							1 **H** hydrogen 1											4 **He** helium 2
7 **Li** lithium 3	9 **Be** beryllium 4												11 **B** boron 5	12 **C** carbon 6	14 **N** nitrogen 7	16 **O** oxygen 8	19 **F** fluorine 9	20 **Ne** neon 10
23 **Na** sodium 11	24 **Mg** magnesium 12												27 **Al** aluminium 13	28 **Si** silicon 14	31 **P** phosphorous 15	32 **S** sulfur 16	35.5 **Cl** chlorine 17	40 **Ar** argon 18
39 **K** potassium 19	40 **Ca** calcium 20	45 **Sc** scandium 21	48 **Ti** titanium 22	51 **V** vanadium 23	52 **Cr** chromium 24	55 **Mn** manganese 25	56 **Fe** iron 26	59 **Co** cobalt 27	59 **Ni** nickel 28	64 **Cu** copper 29	65 **Zn** zinc 30		70 **Ga** gallium 31	73 **Ge** germanium 32	75 **As** arsenic 33	79 **Se** selenium 34	80 **Br** bromine 35	84 **Kr** krypton 36
85 **Rb** rubidium 37	88 **Sr** strontium 38	89 **Y** yttrium 39	91 **Zr** zirconium 40	93 **Nb** niobium 41	96 **Mo** molybdenum 42	[98] **Tc** technetium 43	101 **Ru** ruthenium 44	103 **Rh** rhodium 45	106 **Pd** palladium 46	108 **Ag** silver 47	112 **Cd** cadmium 48		115 **In** indium 49	119 **Sn** tin 50	122 **Sb** antimony 51	128 **Te** tellurium 52	127 **I** iodine 53	131 **Xe** xenon 54
133 **Cs** caesium 55	137 **Ba** barium 56	139 **La*** lanthanum 57	178 **Hf** hafnium 72	181 **Ta** tantalum 73	184 **W** tungsten 74	186 **Re** rhenium 75	190 **Os** osmium 76	192 **Ir** iridium 77	195 **Pt** platinum 78	197 **Au** gold 79	201 **Hg** mercury 80		204 **Tl** thallium 81	207 **Pb** lead 82	209 **Bi** bismuth 83	[209] **Po** polonium 84	[210] **At** astatine 85	[222] **Rn** radon 86
[223] **Fr** francium 87	[226] **Ra** radium 88	[227] **Ac*** actinium 89	[261] **Rf** rutherfordium 104	[262] **Db** dubnium 105	[266] **Sg** seaborgium 106	[264] **Bh** bohrium 107	[267] **Hs** hassium 108	[268] **mt** meitnerium 109	[271] **Ds** darmstadtium 110	[272] **Rg** roentgenium 111								

Elements with atomic numbers 112–116 have been reported but not fully authenticated

* The lanthanoids (atomic numbers 58–71) and the actinoids (atomic numbers 90–103) have been omitted.

Glossary

acid rain rain that has a low pH due to dissolved sulfur dioxide, nitrogen oxides and other impurities

active immunity immunity developed by the body to foreign invading organisms

adapt the characteristics of an organism that make it well suited to living in a particular environment

addictive the need to keep taking a drug

aerobic respiration with oxygen

alkane a family of hydrocarbons containing only single carbon-carbon bonds. Alkanes have a general formula C_nH_{2n+2}

alkene a family of hydrocarbons containing a carbon-carbon double bond. Alkenes have a general formula C_nH_{2n}

alloy a mixture of metals or a metal and carbon (in the case of steel)

alpha a particle emitted by decay of some radioactive materials

alternating current a flow of charge in a circuit which keeps on changing direction

amplitude the distance of a crest of a wave from its rest position

amylase an enzyme that digests starch to maltose

anaerobic respiration without oxygen

analogue a signal which can have any value within a range

antibiotic medicine that kills bacteria

antibody a chemical produced by the immune system to destroy foreign invading organisms in the body

antigen a thing that is foreign to the body

antioxidants substances that slow down the rate of oxidation of food

asexual reproduction reproduction that does not involve fusion of sex cells

asteroid a lump of rock, smaller than a planet, in orbit around the Sun

atmospheric pollution contaminants of the environment that are a by-product of human activity. They include particles (smoke) and gases such as sulfur dioxide

axon the long extended part of a nerve cell

baking powder supplies carbon dioxide during the baking process so cakes will rise. It contains sodium hydrogencarbonate

bases the four chemicals A, C, T and G that code for the instructions for life in DNA

beta a particle emitted by decay of some radioactive materials

Big Bang theory the idea of how the Universe came into existence

binding medium the material that hardens in a paint to form a hard layer. Linseed oil was the original binding medium

binocular-vision using both eyes to view the same object so that distance can be judged more accurately

binomial naming organisms with two names, one for genus and one for species

biodegradable substances that can be broken down by such processes as decomposition by bacteria and can therefore be reused by living organisms

black dwarf the remnant of a star that no longer emits light

black hole remnant of a supernova, so massive that light cannot escape

boiling point temperature at which a liquid rapidly turns to a vapour. Water has a boiling point of 100 °C at normal atmospheric pressure

brain the enlarged front end of the CNS (central nervous system) responsible for much of the coordination in the body

calorimetric experiments involving energy changes that can be measured using temperature changes

camouflage organisms hiding from predators by disguising themselves to look like their surroundings

carbohydrases enzymes that digest carbohydrates

carbohydrate a compound of carbon, hydrogen and oxygen which fits a formula $Cx(H_2O)y$. Glucose, $C_6H_{12}O_6$, is an example of a carbohydrate

carbon dioxide a gas produced by respiration by both plants and animals and then used by plants for photosynthesis

carbon monoxide a compound of one carbon atom and one oxygen atom, CO. It is formed by the incomplete combustion of carbon and carbon compounds. Carbon monoxide is colourless, odourless and poisonous

catalyst a substance that alters the rate of a chemical reaction without being used up

cellulose a substance made by plants that forms the structure of their cell walls

cement a substance made by mixing powdered limestone with clay. When mixed with water it sets to a hard mass

chromosome a structure composed of DNA and found in the nucleus of cells

cilia small hair like structures on the surface of cells

cirrhosis a disease where the liver becomes damaged

climate change changes in the climate such as global warming, brought about by the activities of humans

clone a cell, or group of cells, produced from one ancestor

colloid a state where very small particles of one substance are spread evenly through another

combustion burning of a substance with oxygen to release energy

comet lump of ice and dust in a highly elliptical orbit around the Sun

community all the organisms that live in a particular habitat

compete continual struggle that organisms have with each other for resources

complete combustion when a substance burns in a plentiful supply of air or oxygen to release the maximum amount of energy

composite a material that is made up of other materials

concentration quantity of solute dissolved in a stated volume of solvent

concrete a construction material using cement, sand and aggregate (small stones) mixed with water

conduction how heat energy is transferred through solids

conductor material which allows heat to flow easily

construction materials materials that are used in building

contraceptive pills pills taken to prevent conception

convection how heat energy is transferred by the bulk motion of liquids and gases

core the centre part of the Earth

corrosion the wearing away of the surface of a metal by chemical reaction with air and water

cosmic ray nuclear radiation from space

cracking breaking down of long-chain hydrocarbon molecules by the action of a heated catalyst, or by heat alone, to produce smaller molecules

crater circular depression left in surface by impact of an asteroid

crest the highest point reached by a wave in a cycle

crude oil a mixture of hydrocarbons produced by the action of high temperatures and pressures on the remains of sea creatures over millions of years

crust the outer layer of the Earth

dehydration becoming short of water

diabetes a disease in which a person does not produce enough insulin to control the level of sugar in the blood

diastolic blood pressure when the heart is relaxing

diffusion a movement of particles from an area of high concentration to an area of low concentration

digital a signal which can only have one of two values within a range

dip net a net used to collect organisms from ponds and rivers

direct current electricity due to a flow of charge in just one direction

DNA the molecule that codes for the instructions to make a new living organism

dominant an allele that always expresses itself

drugs chemicals that produce a change in the body

dynamo effect where a voltage is created by a change of magnetism in an electric circuit

earthquake a series of vibrations that shake the Earth's surface, caused by movement of the Earth's tectonic plates

earthquake sudden release of energy in the Earth's crust which generates waves

ecosystem a system of interacting organisms that live in a particular habitat

effector an organ, such as a muscle, that causes a response to a stimulus

efficiency ratio of the useful energy output to the total energy input in an energy transfer process

electrolysis the decomposition of a compound by the passage of electricity

electromagnetic spectrum set of waves propagated as oscillations of an electromagnetic field

elliptical shaped like an egg, a squashed circle

emulsifier (or emulsifying agent) a chemical that coats the surface of droplets of one liquid so they can remain dispersed in the other

endangered organisms that are in danger of becoming extinct

endocrine secreting directly into the blood

endothermic a reaction that takes in energy from the surroundings

energy the ability to do work

E-numbers system for listing permitted food additives. All permitted additives are given an E-number, e.g. E124

enzymes organic catalysts that speed up the rate of a reaction

ester a sweet-smelling liquid formed when an organic acid and an alcohol react. Esters are used in perfumes and food. Methyl ethanoate is an example of an ester

evolution adaptation of organisms to changes in the environment through natural selection

exothermic a reaction which gives out energy to the surroundings

explosion a very rapid reaction accompanied by a rapid release if gaseous products

exponentially an increase that becomes more rapid with time

extinct organisms that no longer exist on the Earth

fat food storage molecules made from fatty acids and glycerol

fertilise fusion of male and female sex cells to form an individual

fertility the ability to fertilise when male and female sex cells fuse together

fibre indigestible plant material found in food

finite resource a resource in limited supply that will run out in the future

food additive a substance added to food to act as a colouring agent, preservative, emulsifier, flavour enhancer etc.

fossil fuel a fuel produced from the slow decay of dead animals and plants and high temperatures and high pressures, found in the ground

fraction a product collected on fractional distillation of crude oil. A fraction has a particular boiling point range and particular uses

fractional distillation method for separating liquids with different boiling points

frequency the number of oscillations of a wave in a second

galaxy a group of many stars held together by their gravity

gamete a cell involved in reproduction, such as an ovum or a sperm

gamma a high frequency wave emitted by decay of some radioactive materials

gene a section of DNA that codes for one specific instruction

generator a device that transfers kinetic energy of a rotating shaft into electrical energy

genetic code the sequence of bases that code for the instructions to make a new living organism

global warming an increase in the Earth's temperature, usually caused by pollution

glucose a type of sugar that is produced by photosynthesis and used as an energy source in respiration

granite an igneous rock formed inside the Earth by crystallisation of molten magma

habitat a place where organisms live with specific environmental conditions

heat energy the energy required to change the temperature of an object

heat-stroke a condition caused when the body overheats

homeostasis maintaining a constant internal environment within the body

hormone a chemical messenger that is released from a gland and travels through the blood

host a live organism that is fed upon by a parasite

hydrocarbon a compound of carbon and hydrogen only

hydrophilic a substance that has a liking or attraction for water

hydrophobic a substance that repels water

hypothermia a lowering of body temperature

igneous rock rock formed when magma crystallises

incomplete combustion when substances burn in an insufficient amount of air or oxygen. It results in less energy release and possibly soot and/or carbon monoxide

indicator species the presence of a species that indicates the quality of the habitat

infrared radiation how heat energy is transferred by infrared waves

insoluble describes a substance that does not dissolve in a solvent

insulation material used to reduce heat energy flow

insulin a hormone produced by the pancreas that lowers the level of glucose in the blood

invertebrates animals without a backbone

ionisation the process of removing or adding electrons to an atom

key a means of identifying different organisms

kilowatt a thousand watts, a thousand joules in each second

kilowatt-hour the energy transferred by an electrical device with a power of one kilowatt over an hour

kinetic energy energy transferred to an object to set it in motion

kwashiorkor a disease caused by a lack of protein in the diet

lactic acid a chemical that causes muscle fatigue and is produced during anaerobic respiration

laser a device which emits electromagnetic waves, all with the same frequency and direction

lava molten rock that escapes through a volcano

limestone a sedimentary rock formed from the remains of sea creatures. It is a form of calcium carbonate

lipase enzyme that digests fat

lithosphere the crust and the uppermost layer of the mantle

longitudinal type of wave which produces oscillations parallel to its direction of travel

LPG (liquefied petroleum gas) the gas which leaves the top of the fractional distillation column and is liquefied. It is an alternative to petrol or diesel in cars

magma the rock between the crust and the core of the Earth

magnetic field the lines of force around an electric current which affect magnetic materials

malleable a material, such as a metal, that can be beaten into thin sheets

mantle a thick layer of dense semi-liquid rock below the Earth's crust

marble a metamorphic rock formed by the action of high temperatures and pressures on limestone. A form of calcium carbonate

microbe a minute living organism

microwave region of the electromagnetic spectrum which lies between infrared and radio waves

minerals essential salts that are required by living organisms

monocular-vision eyes on the side of the head so that each eye looks at a different object giving good all round vision

monomer a small molecule which joins together with other molecules to produce a polymer

Morse code alphanumeric code made from long and short pulses

motor a nerve cell that carries instructions out from the brain

mutation a change to the structure of a gene or DNA caused by such things as chemicals, X-rays or radiation

mutualism a relationship between two organisms of different species in which both organisms benefit

national grid cables which carry electrical energy from power station to consumer

natural dye a material from minerals or plants used to colour fabrics etc.

natural selection a process in which organisms that are most suited to the environment survive and produce more offspring

Near-Earth Object (NEO) asteroid or comet whose orbit crosses that of the Earth

neurone a name for a nerve cell

neutron star very dense remnant of a supernova, not heavy enough to form a black hole

nicotine an addictive chemical that is found in tobacco

non-biodegradable not broken down by the action of bacteria in a landfill site

non-renewable fuel a fuel that took a long time to form and cannot quickly be replaced

nuclear power can be used to transfer nuclear energy into heat energy

nuclear radiation ionising radiations from changes in the nucleus of an atom

oil liquid fat

optical fibre a thin strand of very transparent glass for carrying pulses of light long distances

orbit path followed by an object bound by the gravity of another object

ore a rock that contains a metal or a metal compound in sufficient quantity to make the extraction of the metal economically viable

organism a living animal or plant

ozone a form of oxygen that prevents ultraviolet light from the Sun reaching the Earth

parasite an organism that lives on or in another living organism causing it harm

passive immunity short lasting immunity that is gained by injecting antibodies from a donor

pathogens disease causing organisms

payback time the time for the money saved in heating to equal the installation cost

phosphorescent pigments pigments that store energy when in light and can release this energy again in the dark, and so glow in the dark

photocell a device which transfers light energy into electrical energy

photosynthesis the process by which plants convert water and carbon dioxide into oxygen and glucose using the energy from the Sun

pigment a coloured material used in paints etc.

pit-fall trap a device for catching small animals

planet object which orbits around a star

plutonium a waste material from nuclear power which can be used to make nuclear bombs

pollution the presence in the environment of substances which are harmful to living things

polymer long chain molecule built up of a large number of smaller units, called monomers, joined together by the process of polymerisation

pooter a device for catching small insects

population a group of organisms of one species living together in a habitat

power the energy transferred by a device in one second

predator an animal that hunts and kills other animals for food

prey an animal that is hunted and killed for food by a predator

products substances that are formed in a chemical reaction. They appear to the right of the arrow in a chemical equation

protease enzyme for digesting protein

protein a large molecule made from a combination of amino acids

P-wave longitudinal wave produced by an earthquake

quadrat a square grid used to determine population density

quarry a large unsightly hole in the ground from which useful rocks are removed

radioactive materials substances which emit nuclear radiation

radioactive waste unwanted material which emits ionising radiations

rate of reaction the speed with which products are formed or reactants are used up

reactants substances that are used at the start of a chemical reaction. They appear to the left of the arrow in a chemical equation

receptor an organ or cell that receives an external stimulus

recessive an allele that only expresses itself if the dominant allele is not present

recycled materials that are used again rather than disposed of

red giant star which is fusing helium to form carbon

reflex arc the pathway along which nerve impulses pass in a simple reflex action

refraction the change of direction of a wave which passes from one material to another

rehabilitation the process that enables people to return to a normal life

relay a nerve cell that passes an impulse between two other nerve cells

renewable biomass plant material which can be a source of energy

resources chemicals and materials that can be used for the benefit of humans

respiration a process that takes place in living cells converting glucose and oxygen into water and carbon dioxide with a release of energy

rust a product formed when iron or steel are exposed to air and water. This brown compound is a hydrated iron(III) oxide

satellites objects in orbit around a planet

seismometer a device for recording S-waves and P-waves

sensory a nerve cell that carries information to the brain

sexual reproduction reproduction that involves the fusion of sex cells

sheath the fatty coat that surrounds a nerve cell

sodium hydrogencarbonate the active ingredient in baking powder

solar flares clouds of charged particles ejected from the Sun

soluble describes a substance that dissolves in a solvent

solute the substance that dissolves in a solvent to form a solution

solution what is formed when a solute dissolves in a solvent

solvent a liquid in which a solute dissolves

spacecraft device which can travel in space, away from Earth

species a group of organisms that breeds and produces fertile offspring

specific heat capacity energy required to raise the temperature of one kilogram by one degree

specific latent heat energy required to melt or boil one kilogram

spinal-cord a large collection of neurones that runs up the vertebral column and connects with the brain

star sphere of hot gas which emits light and heat

starch a carbohydrate food storage substance produced by plants

sun protection factor (SPF) how much longer a cream allows you to stay in the sun before burning

supernova explosion of a star when it becomes unstable at the end of its life

surface area the area of the surface of a solid object, usually measured in cm^2

sustainable resources the use of the Earth's resources at a rate at which they can be replaced

sustainable-development development using the Earth's resources at a rate at which they can be replaced

S-wave transverse wave produced by an earthquake

synthetic dye a material manufactured ,usually from coal tar, to colour fabrics etc.

synthetic made by humans, not occurring naturally

systolic blood pressure when the heart is contracting

tectonic plates very large plates of rock which float and move very slowly on the mantle of the Earth

temperature a measure of how hot or cold an object is

thermal decomposition decomposition of a compound by heating

thermochromic pigments pigments that can change colour at different temperatures

tolerant when the body becomes resistant to a particular drug or chemical

total energy input the energy transferred into a device for it to accomplish its task

total internal reflection reflection inside a material when the angle of incidence exceeds the critical angle

toxicity a measure of how poisonous a substance is

toxins chemicals produced by microorganisms that damage the body

transformer a device which raises or lowers the voltage of an alternating current

transmission frequencies range of frequencies allocated to a TV or radio channel

transverse wave type of wave which produces oscillations at right angles to its direction of travel

trough the lowest point reached by a wave in a cycle

ultraviolet region of the electromagnetic spectrum between visible and X-rays

uranium an element used as a fuel in nuclear reactor

useful energy output the energy transferred by a device to do some task

vectors organisms that transmit and carry a disease

vertebrates animals that have a backbone

vitamins essential chemicals that are required by living organisms

volcano a tube from inside the Earth to its surface through which lava can escape in an eruption

water a molecule that is essential to all living organisms

watt the unit of power, one joule per second

wavelength the distance between adjacent crests of a wave

white dwarf the remnant of a star that has run out of fuel for fusion reactions

wind turbine a device which transfers kinetic energy of wind into electrical energy

withdrawal symptoms symptoms that appear when a person stops taking a drug that they are addicted to

Index

Revision Guides

Beat the rest - exam success with Heinemann

Ideal for homework and revision exercises, these differentiated **Revision Guides** contain everything needed for exam success.

- Summary of each item at the start of each section

- Personalised learning activities enable students to review what they have learnt

- Advise from examiners on common pitfalls and how to avoid them

Please quote S 603 SCI A when ordering

(t) 01865 888068 (f) 01865 314029 (e) orders@heinemann.co.uk (w) www.heinemann.co.uk

L554